Alexander B. Macdowall

Weather and Disease

A Curve History of their Variations in Recent Years

Alexander B. MacDowall

Weather and Disease
A Curve History of their Variations in Recent Years

ISBN/EAN: 9783337336998

Printed in Europe, USA, Canada, Australia, Japan

Cover: Foto ©berggeist007 / pixelio.de

More available books at **www.hansebooks.com**

Weather and Disease:

A CURVE HISTORY

OF THEIR

VARIATIONS IN RECENT YEARS.

BY

ALEX. B. MacDOWALL,

M.A., F. R. Met. S.

LONDON:

THE GRAPHOTONE CO., 1 MACLEAN'S BUILDINGS, E.C.

1895.

PREFACE.

THE influence of weather on health, a matter which concerns us every moment of our lives, is yet one that is greatly in need of elucidation. With growing attention to it, we become aware how little it is understood, and how many unsolved problems it presents. There are signs, here and there, of a deepening interest in this important subject ; thus, at Rome, *e.g.*, a meteorological station has been recently attached to the laboratories of the Public Health Department, and students of the annexed school will be instructed in the use of the instruments, and attend a course of lectures on Meteorology applied to Hygiene.

The primary object of this book is to give an idea of the way in which certain elements of our weather, and the mortality from some well-known diseases, have varied in recent years. While some direct attempt is made to trace the relations between weather and disease, the work done is, generally, rather in the way of furnishing data for comparison and study. Some of the facts presented, which have been met with in going along rather unfrequented paths, may be found, the author believes, both new and instructive.

The mode of exposition adopted is largely that of graphic curves ; these, indeed, may be said to be the essence of the book. What is written about the charts may be soon read. The reader is invited to study them independently, and to follow out, with care, any trains of thought which they may perhaps suggest to him.

It appears to the writer that the graphic method is far too little utilised. In the best serials and books dealing with figures, one may find long and difficult accounts of how this or that item has varied in a series of years, or other given time. A few simple lines on a cross ruled space would at once make the whole thing clear. In the actual case it is like groping for something in a dark room, where one might first turn on the light of a glow lamp.

Columns and rows of figures are often, of course, indispensable. But sometimes they are not. It may be desired, in surveying a wide domain, merely to point out some salient features here and there ; to indicate some noteworthy recurrence, the correspondence in phase of two or more curves, or the like. Results may be sought, rather than processes ; and a few figures may be given, while the others may be left to be guessed approximately from the curves. The graphic chart, indeed, is a *multum in parvo* mode of information, which saves trouble in description, and presents the truth of figures concisely, clearly, and impressively. It is remarkable, moreover what new points of view are often gained when a column of figures is converted into a curve.

A greatly extended use of the method, therefore, is here advocated, and it is rendered practicable by the greatly enlarged facilities of printing. Even a rough diagram may be regarded as better than none. Those here offered make no pretention to fineness of quality or mathematical precision.

Many of these curves are said to be *smoothed*. It may be well to explain this.

The object of smoothing is to bring out those longer waves of variation which may be dimly perceived through the zigzags of the actual curve, like a long ocean swell, obscured somewhat by the minor waves and ripples. Suppose we have a series of annual or other values, a, b, c, d, e, f, g, &c., and we want to smooth these with averages of 5. We first take the average of a to e, and put it down in place of c ; next the average of b to f, and put it down in place of d ; and so on. The averages chiefly used in the charts are those of 5 and 10. In the latter case, the result is put down in the fifth place. Sometimes the smoothed curves are given without those of the actual figures ; but in many other cases both are given, so that the nature of the relation between them need not be misunderstood.

The existence of demonstrable relations between weather and the sunspot cycle of about eleven years, is now regarded by many meteorologists as highly probable ; considering, especially, the proved relations of sunspots to magnetic phenomena and to auroras, and the obvious influences of solar radiation. The notion that a given sunspot phase (maximum or minimum), if it affect weather, must affect it everywhere alike (giving everywhere excessive rainfall, excessive heat, or other character), now appears to be untenable. If so, the value of evidence from individual stations is enhanced, and it seems desirable that any series of weather variations noticed to present some fair amount of correspondence with the sunspot cycle through a considerable number of its wave periods, should be brought forward and discussed. It is in this spirit, and entirely without dogmatism, that the four diagrams relating to sunspots and weather are offered for consideration.

Some of the weather curves in this collection have previously appeared in *Nature*, and are here reproduced by kind permission of the Editor. The author's thanks are also due to Mr. F. J. BRODIE and Mr. T. H. BAKER for the extension, to date, of useful tables published by them ; one relating to fog, the other to harvests. In preparing the section on disease, STEVENSON & MURPHY's excellent *Treatise on Hygiene and Public Health* has been found helpful.

It would be sanguine, perhaps, to imagine that in dealing with so large a mass of figures, and making so many calculations, one has never failed in accuracy. Should any mistake come to notice, the author will be glad to have it pointed out. He believes the degree of care exercised has been such as at least to avoid any grave error.

CONTENTS.

I.—WEATHER.

II.—DISEASE.

8

SHEPHERD : " Weel, do ye ken, sir, that I never saw in a' my born days what I could wi' a safe conscience hae ca'd bad weather ? The warst had aye some redeemin' quality aboot it that enabled me to thole it withoot yaumerin'. Though we may na be able to see, we can aye think o' the clear blue lift. Weather, sir, aiblins no to speak very scientially in the way o' meteorological observation,—but rather in a poetical, that is, a religious spirit,—may be defined, I jalouse, 'The expression o' the fluctuations and modifications of feeling in the heart o' the heevens made audible and visible and tangible on their face and bosom.' That's weather."— *Wilson's* "*Noctes Ambrosianæ.*"

I.—WEATHER.

GENERAL DIRECTIONS.
(Applicable to both Sections)

THE plan adopted with each diagram is, first, to point out briefly what the curves refer to, and then add a few remarks on their sense and bearings. The scales, both vertical and horizontal, are varied according to convenience. Sometimes curves are brought close together, or superposed, having separate vertical scales; in which case a letter is used to connect each curve with its proper scale. A frequently used method is to give the actual variations as a dotted curve, and the result of smoothing as a continuous curve traversing it. References to books, &c., are given at foot.

I.—THE RAINFALL OF ENGLAND.

DIAGRAM : Dotted curve, fluctuations of our annual rainfall since 1800. Continuous curve, the same smoothed with averages of 10.

The numbers here used are not so many inches of rain, but the ratio numbers adopted by Mr. G. J. Symons, our chief rainfall authority. The average yearly rainfall of a place being considered as 100, then if a year, *e.g.*, had rainfall 10 per cent. *over* the average its number would be 110 ; if 10 per cent. *under* average, it would be 90, and so on. The series for the whole country is worked out from the data for a number of different stations.

The actual variations (in those ratios) show great irregularity ; ranging from the very wet year 1852 (ratio 136) to the very dry one 1887 (ratio 69). Several other very wet and very dry years are indicated.

It has been pointed out by Mr. Symons that the years 1834, 1844, 1854, 1864, 1874, and 1884 were all dry years ; but 1894 probably breaks the series. The years about 1878 are remarkable ; we have there 9 successive wet years (1875-1883). Going back to the next marked wave crest in the smoothed curve, 52 years before, we have 5 successive wet years (1827-1831). We are able to go back still further, to the early part of last century ; and our smoothed curve has another notable crest about 1771 (*i.e.*, 55 years back from 1826). Then we have definite minima, or times of drought, about 1745, then about 1804 (59 years interval), and again about 1853 (49 years); showing a certain regularity in these long variations. In the noteworthy group of years 1738-62 only one was above the average. At the beginning of this century there were 10 dry years in succession (1800-9) at the last minimum only 5 (1854-58). Since 1878 the smoothed curve has been in rapid descent, but its future course is not easy to predict. Perhaps we may have a recurrence of prolonged drought like that in last century. (See Symons' *British Rainfall*, 1870, 1881, 1892.)

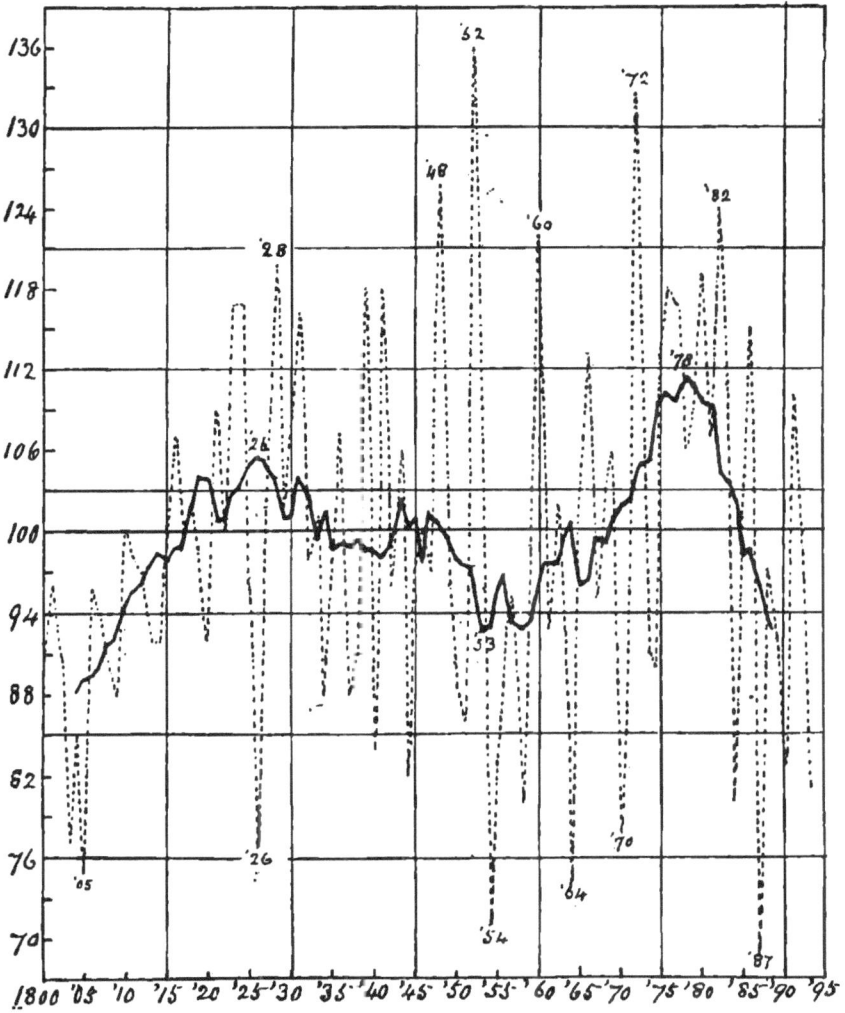

2.—THE WETTEST AND THE DRIEST MONTH.

DIAGRAM : Upper dotted curve, monthly rainfall at Camden Square, London, in 1888-93 (in inches). Continuous curve, result of smoothing with averages of 5 months. (*a*) below, smoothed curve (5 av.) of monthly rainfall in S. W. Ireland (3 stations) ; (*b*) ditto, E. of England (3 stations).

On an average, our wettest month in London is October ; our rainfall is mainly autumnal. But we often, of course, find some other month to be the wettest. In the upper part of the diagram October comes out wettest in 4 of those years ; but July in 1888 and 1890.

The driest month, on an average, is February ; and there is one very dry February in this series, that in 1891 ('01 inch). But it is the driest month in only 2 of the 6 cases (January, September, and April being driest in the others).

In the west of Ireland and north of Scotland there is a different *régime*. The bulk of the rainfall is generally later, in winter. The continuous curve (*a*), from averages of three stations in the Cork region, (Dunmanway, Cork, and Kenmare), lags, in its crests and hollows, behind the London curve, culminating somewhere from October to January. The east of England curve (*b*), from averages of 3 stations (Geldeston, Swaffham, and Holt), resembles the London curve, with autumnal rainfall. On the Continent, north of the Alps, we should find a preponderating summer rainfall. (Symons' *British Rainfall.*)

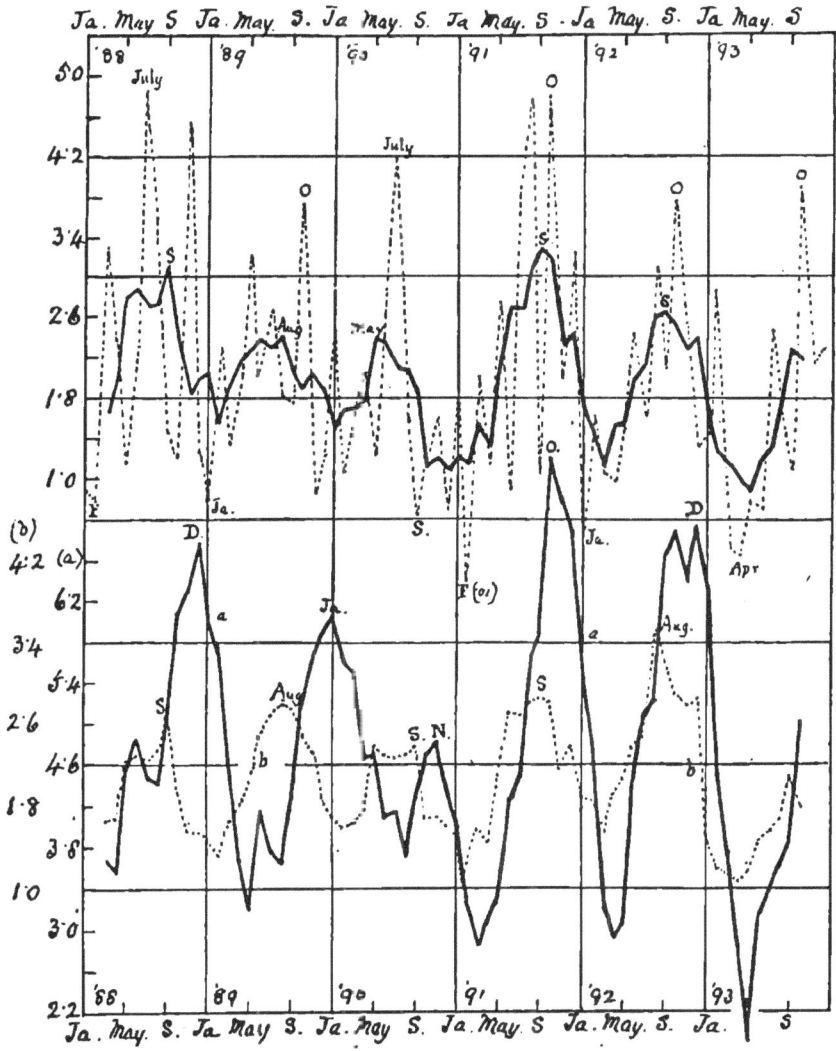

3.—THE BEST TEN DAYS IN SUMMER.

DIAGRAM : Dotted curve, variation of number of years, out of 45, each day of summer (considered separately) has had rain. Continuous curve, smoothed with averages of 10.

This diagram is the result of an examination of the last 45 years' records of the weather at Greenwich in summer (June, July, August), day by day, noting, in the case of each day, how often, that is, in how many of those years, it had rain. By this mode of reckoning a distinctly good time comes out in the end of June ; the continuous curve going down to 14·7 on the 26th and 27th (10 days average from 22nd or 23rd).

It need hardly be said, however, that no one who is " weatherwise " would build immoderate hopes on this or any other part of summer !

The driest day thus reckoned is August 11th, with the value 9 ; while in 4 cases the curve rises to 24. (*Greenwich Observations.*)

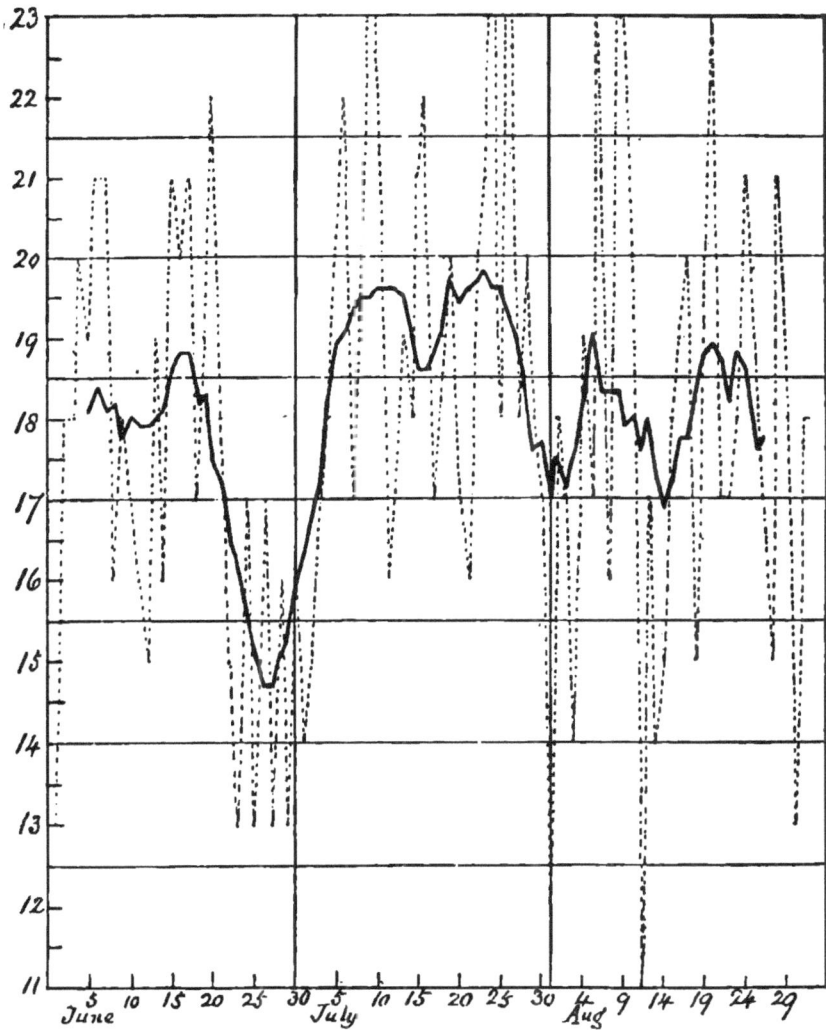

4.—THE WINTERS OF THIS CENTURY.

DIAGRAM : Dotted curve, variation of mean temperature of winter (December to February) at Greenwich since 1800. Continuous curve, smoothed with averages of 10. (Winters are designated by the year in which they end ; thus 1814 means 1813-14.)

There are many ways of estimating winters in respect of severity. That of mean temperature (just indicated) gives a fair measure.

It is difficult here to make out a decided periodicity, such as might enable one to forecast coming winters. The severest winter of the century, so far, appears to have been that of 1813-14 (M. T. 32·5°), when a fair was held on the Thames. The cold was ushered in by one of the worst of London fogs, lasting a week, and this was followed by a snowstorm of 48 hours. There seems to have been a general decrease of severity in winters (increase of mean temperature) till about the sixties, and an increase of severity since. This is indicated by the smoothed curve, also by the series of severe winters we have marked with dots, viz. : 1830, 1841, 1847, 1855, 1860 (ascending series); 1865, 1871, 1879, 1891 (descending series). The M. T. of 1890-91 was 34·0°, and that of last winter, 1894-95, is a little higher (34·8°). The winter of 1854-55 is remembered as the Crimean winter, and that of 1870-71 in connection with the Franco-Prussian war. In December, 1879, the absolutely lowest record of temperature in the United Kingdom was made, viz. :—23°, or 55° below freezing point, at Blackadder, in Berwickshire. Some very mild winters are indicated. In 1868-69 the M. T. was 44·4° ; in 1833-34 and 1876-77 it was 43·4° ; in 1845-46 it was 43·2°. We seem to have come to a time of (at intervals) very severe winters. (See Table of Monthly Temperatures, by Buchan, in *Journal of the Scottish Meteorological Society*, 3rd Series, No. 9.)

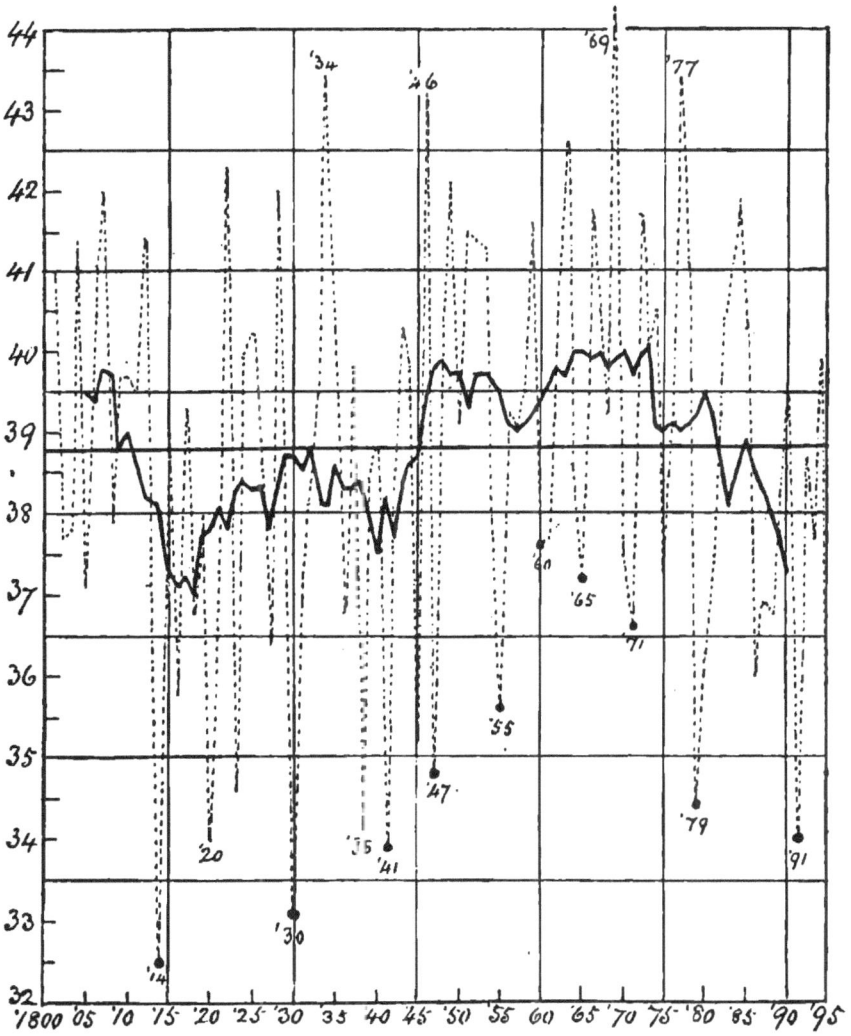

5.—ANOTHER VIEW OF OUR WINTERS.

DIAGRAM : Dotted curve, variations in the number of frost days in October to April, since 1842. Continuous curve, smoothed with averages of 5. (1842 means 1841-42, &c.) The curves are *inverted*. We have considered the mean temperature of the three months, December—February ; we here consider the number of frost days in the seven months, October—April. The vertical scale here increases downwards, so that low points mean severe winters as in the former diagram.

This way of looking at winters leads to some rearrangement of them as regards intensity. Thus the winter season 1887-88 comes into prominence, exceeding both 1878-79 and 1890-91. While it had fewer frosts (and higher mean temperature) than these in the three winter months, it had considerably more frost in the four other months ; March and April were both very cold. Several other cases may be noticed. The number of frost days ranges from 90 in 1887-88 to 24 in 1883-84. (See Harding's Table, *Quarterly Journal of the Royal Meteorological Society*, 1891, p. 113.)

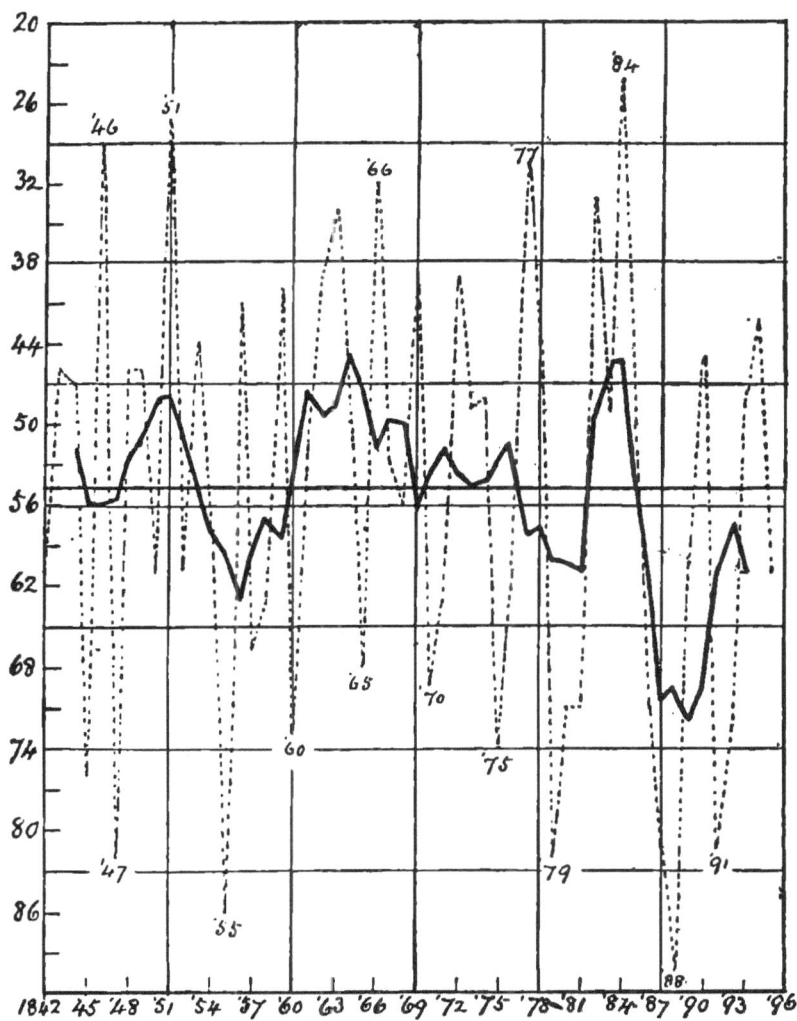

6.—THE OCCURRENCE OF VERY COLD DAYS.

DIAGRAM : Upper curve, variation in number of days in each winter season (at Greenwich, since 1842) with maximum temperature not over 32°. Lower curve, variation in number of days in which the minimum fell below 20°.

A day may fairly be considered very cold, in which the maximum temperature does not rise above freezing point. In 10 of the winters considered there were no such days, and in one case (1890-91) the number rose to 27. The average is about 5. Conspicuous points of the curve occur as follows :—

1844-45	1854-55	1866-67	1878-79	1890-91
13	15	12	18	27

which shows a certain regularity of recurrence at 10 or 12 years' interval. But we cannot say whether this will continue. Last winter (1894-95) had 17 of those days.

The lower curve is drawn from a table by Mr. Harding (extended); it relates to days with minimum under 20°. The Crimean winter (1854-55) here comes into prominence. The last few winters present an unusual succession of high numbers. In 1894-5 there were 11 of those days, all in February. (See *Nature*, vol. 51, p. 416 ; *Greenwich Observations* ; and Harding, *loc. cit.*)

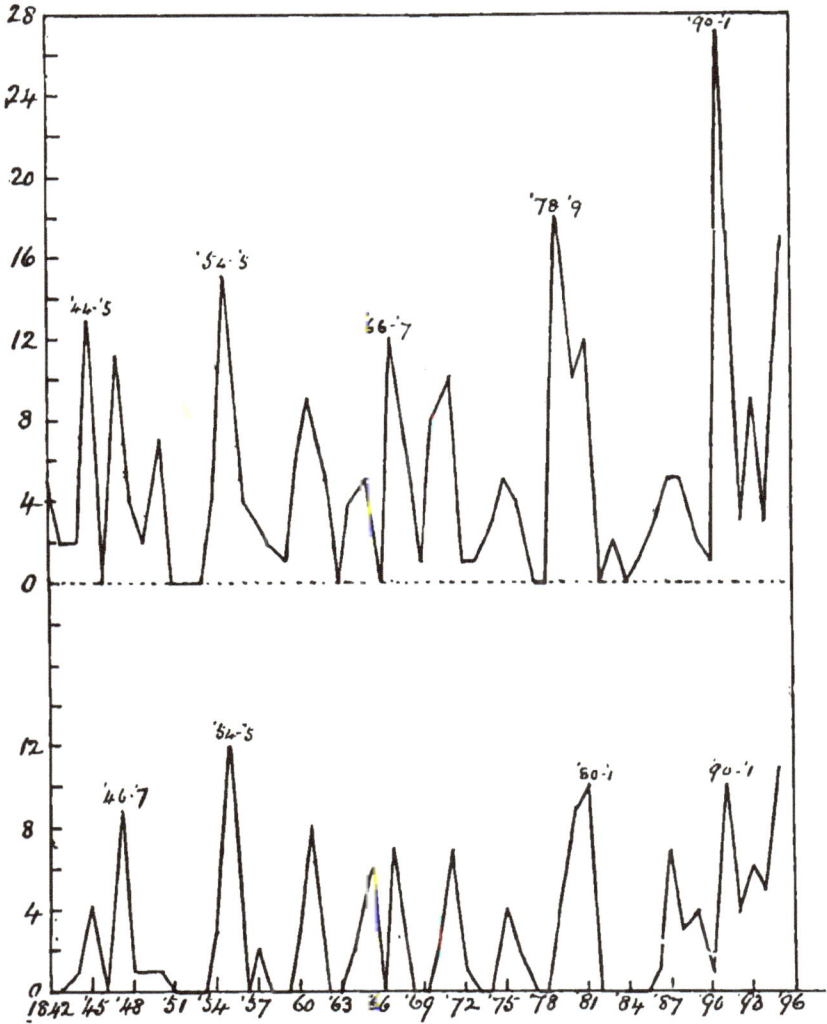

7.—LONG FROSTS.

DIAGRAM : Lines showing the longest continuous frost in each winter, at Greenwich, since 1850-51.

This is in five sections, each representing a month (November—March). The longest continuous frost is that of last winter 1894-95, extending from January 21st to February 20th, *i.e.*, 31 days. There was one of 28 days in February—March, 1886, and one of 26 days in December—January, 1890-91. None in the earlier part of the period come up to these. In the mild winter 1877-78 no frost lasted more than 4 days, and there were 5 groups of that length. (The data are from a table by Mr. Harding, extended.) (*Quarterly Journal of the Royal Meteorological Society*, 1886, p. 235 ; *Greenwich Observations*.)

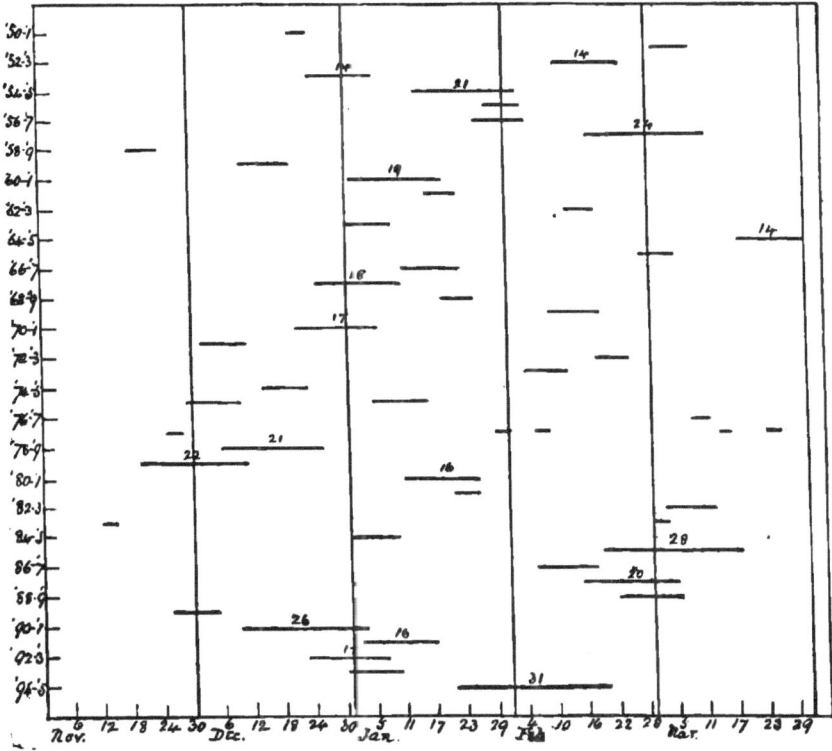

8.—A LONG WAVE OF TEMPERATURE.

DIAGRAM : (*a*) Dotted curve, a smoothed one ($\frac{5}{5}$ av.) of the mean temperature of the first, plus the last, quarter of each year, at Greenwich, since 1830. Continuous curve, averages smoothed in the same way. (*b*) Decade averages of the same mean temperature, beginning with 1830-39 (the last point only 1890-93).

To take another standpoint with regard to the fluctuations of cold, we here consider the mean temperature of the first plus the last quarter in each year. Note, in this dotted curve (*a*), the long rise from a low point in 1838 (40·5°), and, after some fluctuations at a higher level, the return to a low point in 1887 (40·6°). The twice-smoothed curve presents crests at about 10 years' intervals (from 1850).

The lower curve (*b*), giving averages in decades on a larger scale, rises to a maximum in the sixties, and then falls to a point lower than the initial one. (See Buchan's Table, *loc. cit.*)

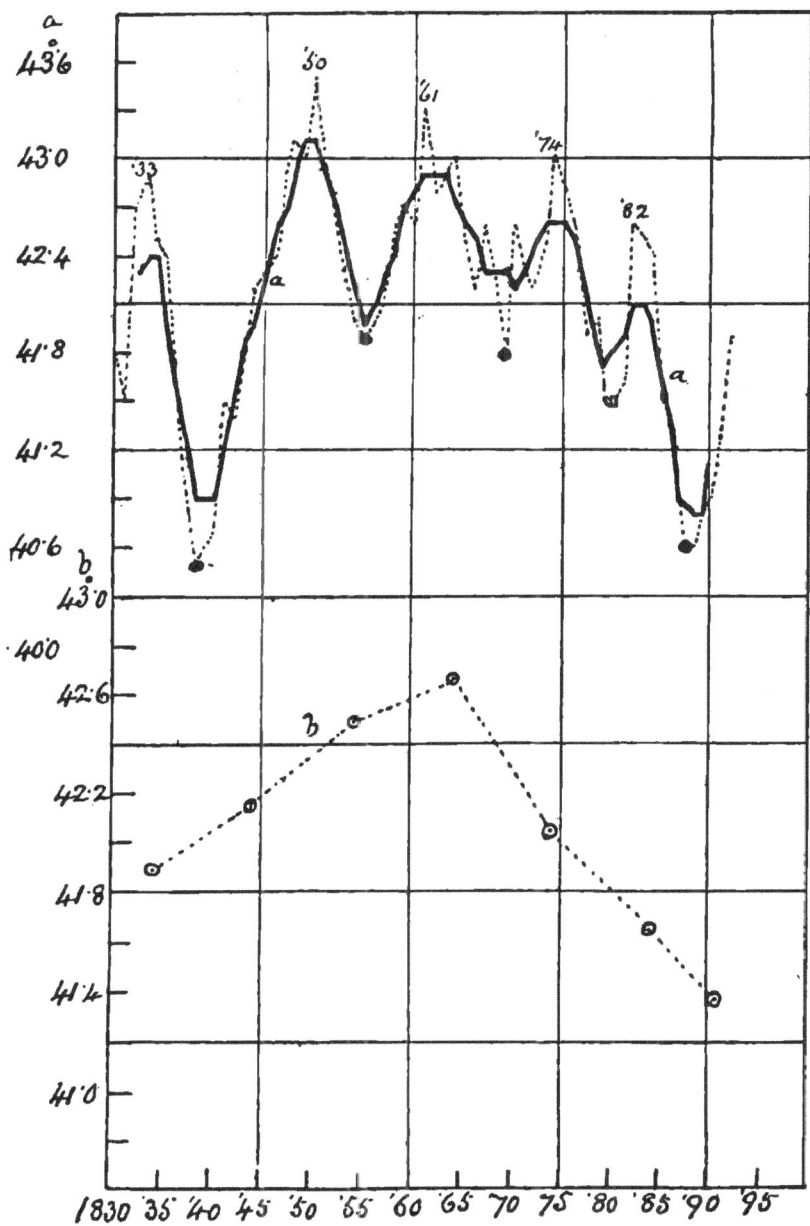

9.—A REMARKABLE WINTER (1890-91).

DIAGRAM : (*a*) Smoothed curve (5 av.) of daily minimum temperature in winter months at Greenwich. (*b*) Ditto at Nairn in the north of Scotland.

The severity of the winter 1890-91 is well remembered in London. Among other effects nearly eight weeks skating was enjoyed.

Were these two smoothed curves of daily minimum temperature put before someone to say which was the London curve and which the Nairn one, he would probably assign the milder to London. This winter, however, was considerably colder in London than in the North.

This (smoothed) London curve goes below freezing point on December 9th, and does not return to it till January 23rd.

An interesting feature appears in those minor waves of variation. They largely correspond in the two curves, those of the London curve lagging, however, a few days behind those of the Nairn curve. This extends through most of the winter. Such lagging correspondence between southern and northern stations may be traced in other cases of severe weather, but not always. (*Daily Weather Report.*)

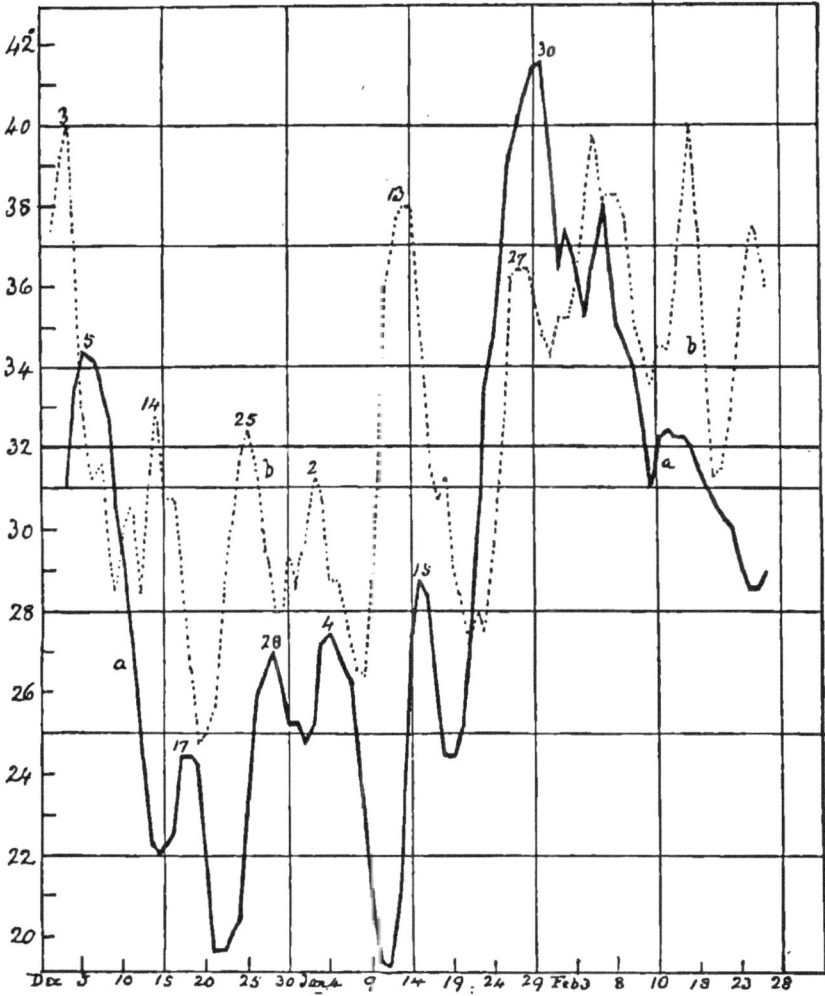

10.—THE SUMMERS OF THIS CENTURY.

DIAGRAM : Dotted curve, variations in mean temperature of summer (June, July, August) at Greenwich since 1800. Continuous curve, twice smoothed with averages of 5.

This chart is similar in design to that of winters, except that in smoothing the actual values are first smoothed with averages of 5, then the averages are similarly smoothed.

The distinctly cold time about 1814 seems to have affected our summers as well as our winters. In the smoothed curve we may detect a certain regularity, with intervals of about 10 years, for a considerable period. But it must be admitted that, as in the case of winters, the actual variations are very irregular.

A few of the very hot summers are indicated. 1846 comes out the highest by this reckoning, with M. T. 66·8°. It was preceded by a backward spring, and was very dry (" hottest June on record "). The next, 1826 (M. T. 66.3°), was also very dry throughout, with a remarkably early and good harvest. 1859 and 1868 may also be noted. No recent summers come up to these. On the other hand we have the very cold and wet summer of 1816, with one of the worst harvests known. " After St. Swithin's day it rained 25 days out of 30." 1860 was another disastrous summer, June wet throughout, also August ; " a wetter season than 1799 and 1816, but not quite so late."

The long wave of variations extending from 1860 to (say) 1879 (another cold summer), may be noted as interesting. (See Buchan's Table, *loc. cit.*, Baker's *Records of the Seasons*).

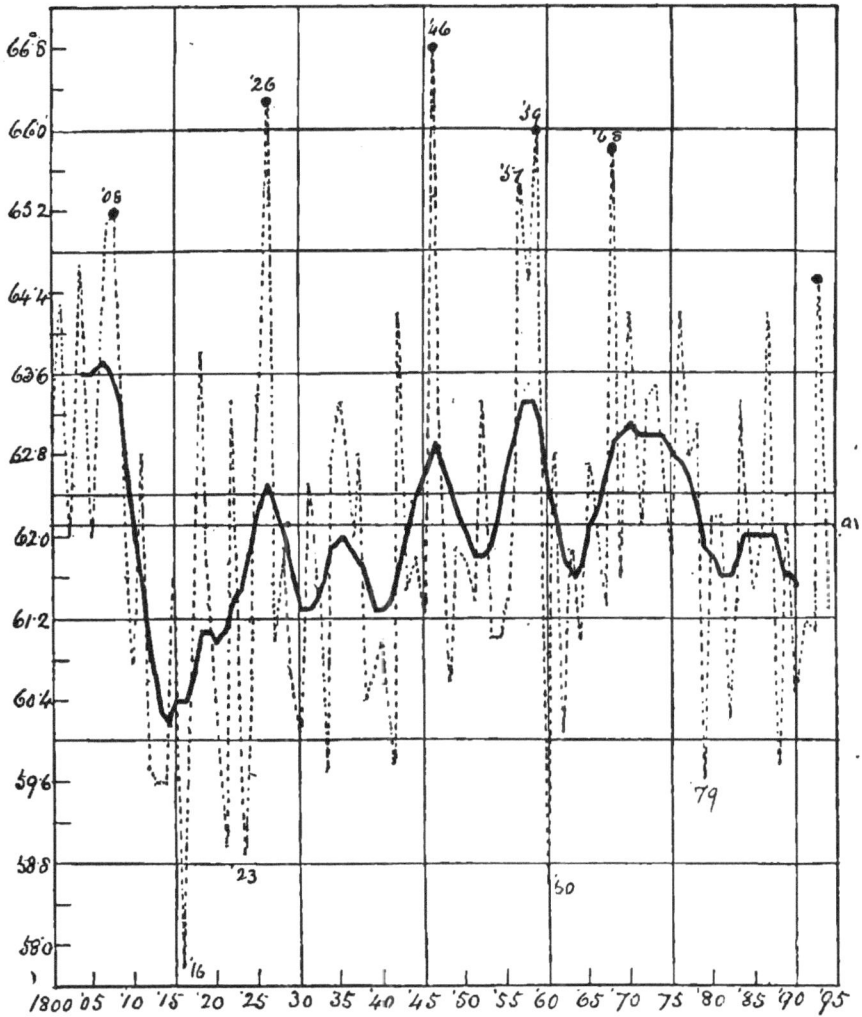

11.—SPELLS OF HEAT AND THEIR PROGRESSION.

DIAGRAM : (*a*) Dotted curve, variations of daily maximum temperature at Parsonstown in August, 1893. Continuous curve, smoothed with averages of 5. (*b*) Ditto for Paris. (*c*) Ditto for Munich.

The spring and summer of 1893 were remarkable in many ways. We here consider the intense heat in August.

These three pairs of curves of maximum daily temperature for Parsonstown (in the heart of Ireland), Paris, and Munich, culminate one after the other in the order given. Thus, the absolute maxima were reached at Parsonstown on the 14th-16th, at Paris on the 18th, and at Munich on the 23rd; the maxima of the smoothed curves on the 14th, 16th, and 21st respectively.

The fall of temperature in this month at some stations was very great and sudden. Thus in Paris we find a maximum of 96° on the 18th, and one of 71° next day.

The kind of progression indicated may often be noticed in comparing spells of heat at various Continental stations and those at our own, and spells of cold show it also. (*Daily Weather Report.*)

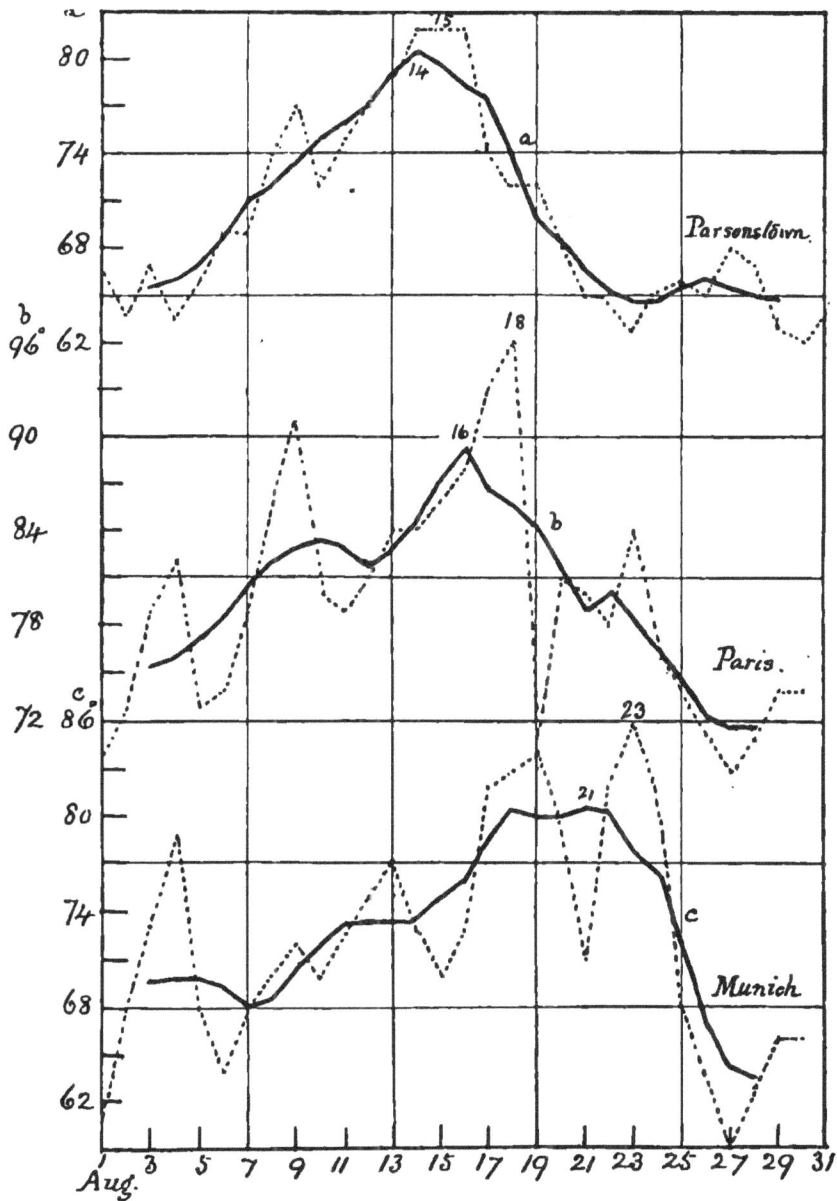

a

80

74

68

b
96° 62

90

84

78

72 86

80

74

68

62

15

14

a

Parsonstown

18

16

b

Paris

23

21

c

Munich

Aug. 3 5 7 9 11 13 15 17 19 21 23 25 27 29 31

12.—BRIGHT SUNSHINE IN SPRING, SUMMER, AND AUTUMN.

DIAGRAM : (*a*) Dotted curve, variations of bright sunshine at Greenwich in spring. Continuous curve, smoothed with averages of 5. (*b*) Ditto for summer. (*c*) Ditto for autumn (with larger scale).

The records of sunshine at Greenwich, now extending over 18 years, present various points of interest. In these three pairs of seasonal curves for spring (March, April, May), summer (June, July, August), and autumn (September, October, November), we may note this relation : We have first a wave crest in the spring, then a few years later, one in the summer, then a few years later, one in the autumn. The actual maximum years are spring 1882, summer 1887, and autumn 1891.

The time is too short, of course, to enable us to say whether there is anything like a cycle in bright sunshine ; but it may be well to notice whether this succession is repeated in future years, and whether the crests of a seasonal wave recur at about 10 years' interval.

The sunshine of the summer of 1887 (Jubilee year) was in sharp contrast to that of the following summer. One was the highest amount in this series (715 hours), and the other the lowest (372 hours). The brightest spring (1893) had 572 hours, the brightest autumn (1891) had 295 hours.

Sunshine observations at Greenwich are made with the Campbell Stokes instrument, in which the sun's heat is focussed on a strip of millboard by a glass globe. (*Greenwich Observations.*)

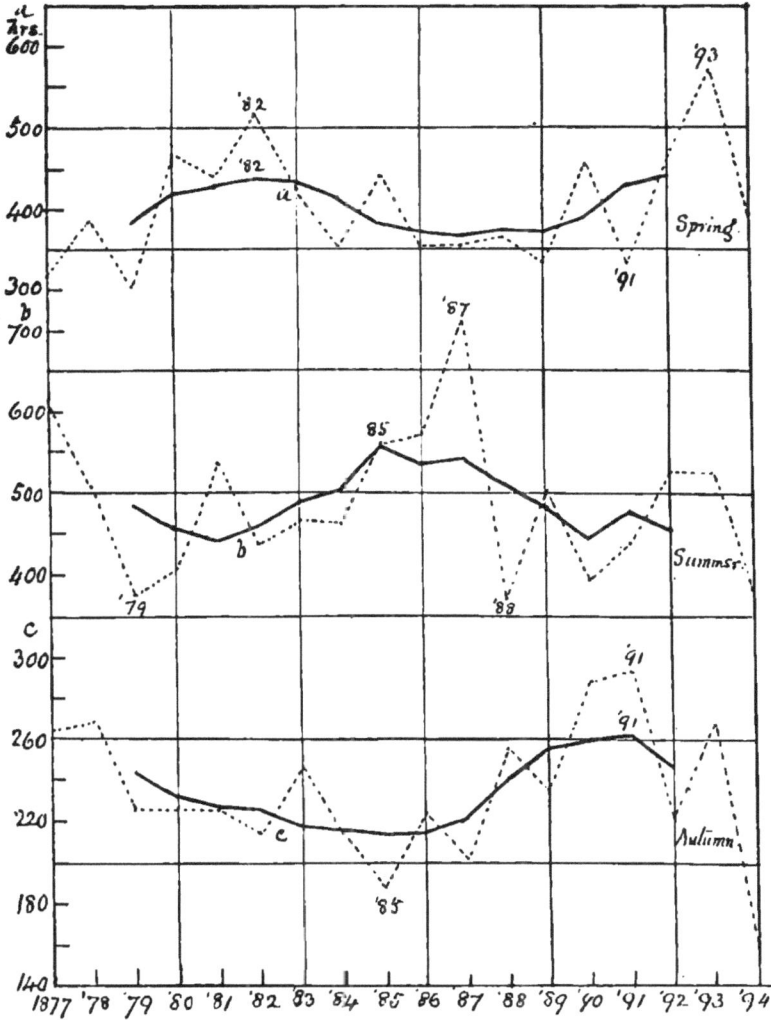

13.—BRIGHT SUNSHINE IN THE WINTER HALF.

DIAGRAM : (*a*) Variations in amount of bright sunshine in October–March at Hillington, Norfolk (from 1881-82). (*b*) Ditto at Cullompton, Devon. (*c*) Ditto at Eastbourne (from 1886-87). (*d*) Ditto at Newton Reigny, Cumberland (from 1884-85). (In chart, 1882 means 1881-82, &c.)

This diagram shows how sunshine in the winter half has recently varied at 4 stations widely apart. It reveals a distinct upward tendency in the amounts of late years. Thus at Eastbourne, the sunshine continuously increased year by year from 381 hours in 1887-88 to 625 hours in 1893-94. Last winter breaks the succession. We may give the Cullompton figures since 1883-84.

	1884	1885	1886	1887	1888	1889	1890	1891	1892	1893	1894
	345	293	351	434	390	384	373	489	494	534	555
Smoothed (3 av.)	—	330	359	392	403	382	415	452	506	528	—

Important effects are doubtless associated with these long variations. (*Meteorological Record.*)

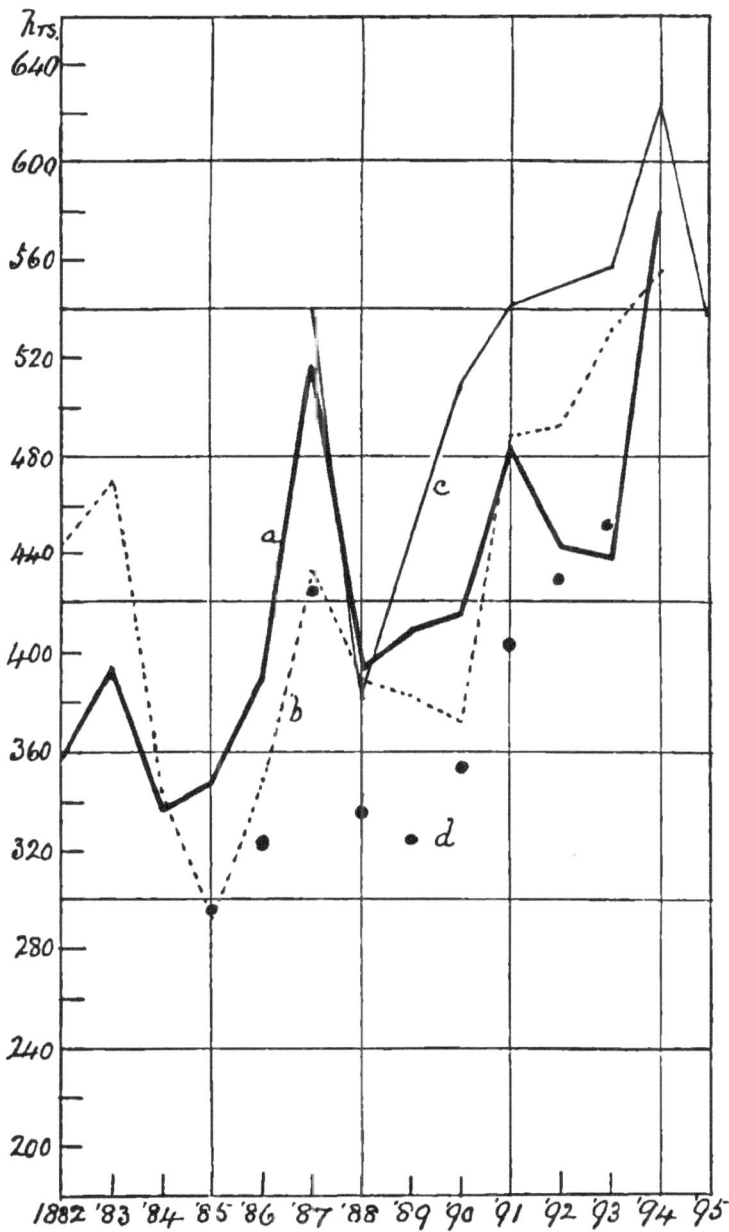

14.—SUNSPOTS AND THUNDERSTORMS.

DIAGRAM : (*a*) Dotted curve, variation in thunder-days at Greenwich (April–October of each year since 1850). Continuous curve, smoothed with averages of 5. (*b*) Similarly smoothed curve of days with thunder at Berlin (in year). (*c*) Ditto for Geneva. (*d*) Sunspot curve (inverted).

We may now proceed to consider a few cases of apparent correspondence between certain weather phenomena and the sunspot curve. Whether they indicate some causal relation, or are merely of fortuitous nature, is another question. In any case it will be interesting to observe how far, in the future, the correspondence is maintained.

We begin with thunderstorms. Our records of this class of phenomena seem rather inadequate. The Greenwich curve (dotted one, *a*) is arrived at by picking out and counting the number of days, in April to October of each year, on which thunder was heard. In the inverted sunspot curve, at foot, there are minima in the years 1856, 1867, 1878, and 1889. Corresponding fairly with three of these we have in the Greenwich smoothed curve maxima in 1857, 1878, and 1890, but no well-marked wave for the sunspot minimum of 1867.

In the similarly smoothed curves for Berlin and Geneva (*b* and *c*), the reader will easily trace by the figures a considerable correspondence of the same kind with the sunspot curve ; maxima of thunderstorms in general following minima of sunspots, and minima of thunderstorms maxima of sunspots. (See *Nature*, Vol. 46, p. 488.)

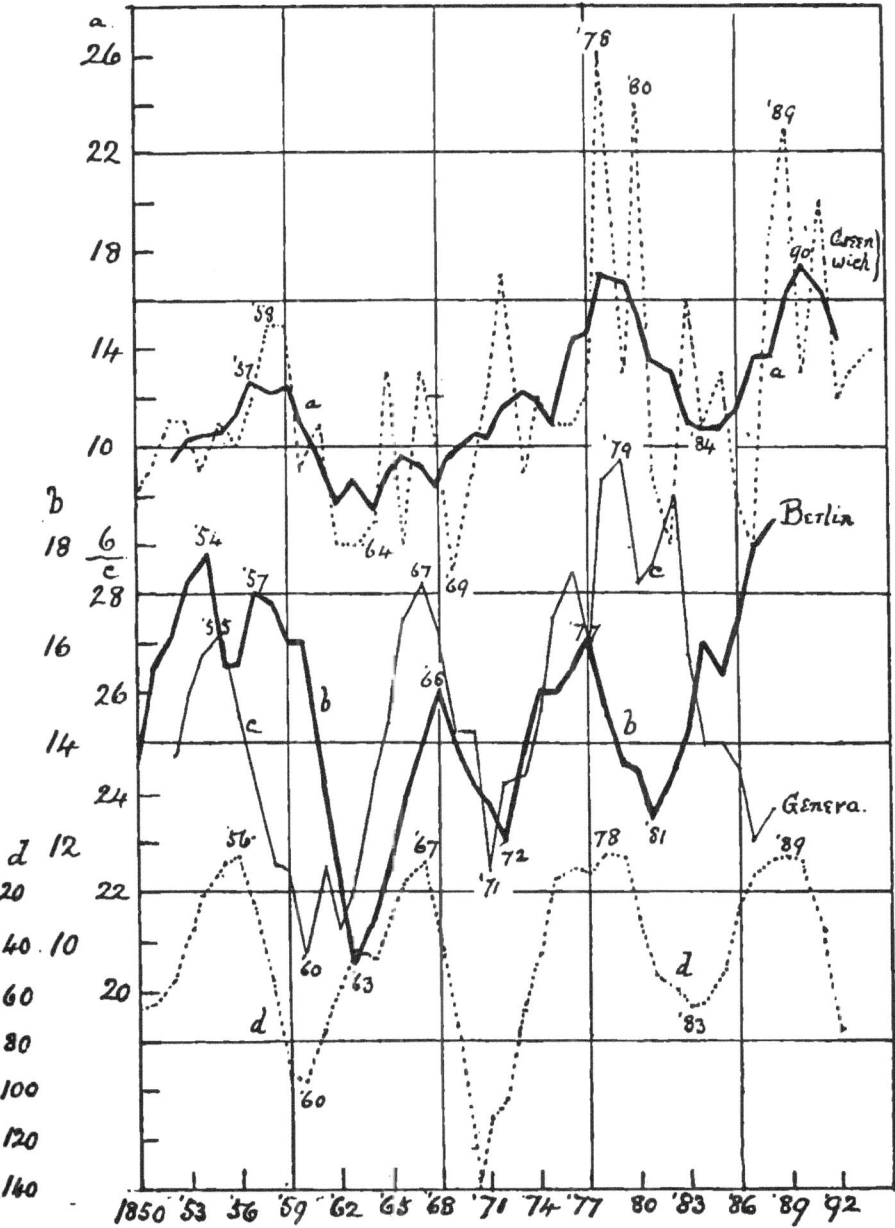

15.—SUNSPOTS AND THE RELATION OF RAINDAYS TO RAINFALL.

DIAGRAM : (*a*) Dotted curve, variation of the ratio of raindays to rainfall at Greenwich each year since 1835. Continuous curve, smoothed with averages of 5. (*b*) Sunspot curve (normal).

If we take the total number of days with rain in a year, and divide by the rainfall of the year, we get the number of raindays per inch of rain. The smoothed curve (for Greenwich) since 1835 presents maxima as follows :

1839	1849	1862	1872	1884

while the sunspot curve shows maxima :

1837	1848	1860	1870	1883

The former maxima, therefore, slightly lag behind the latter (which might be considered natural if it were a case of cause and effect). Similar relations may be noticed between the minima.

According to this view there would appear to be, on the whole, most rain per rainday (or fewest raindays per inch) near or soon after minimum sunspots, and the opposite near or soon after maxima. (*Annual Summary.*)

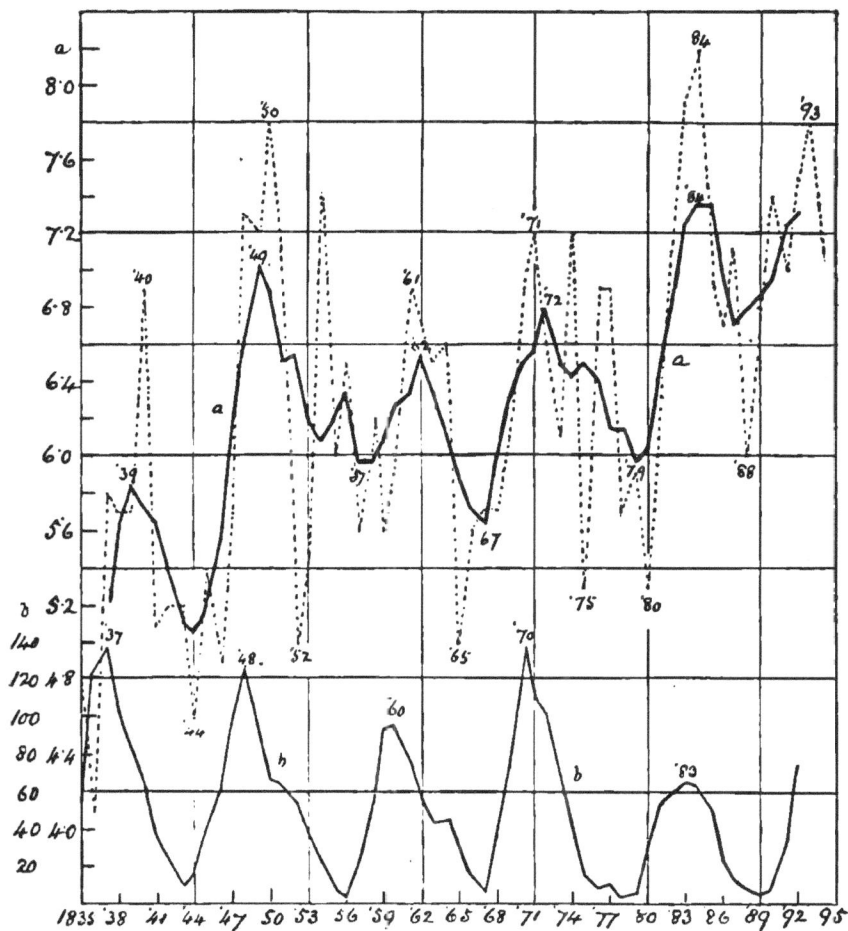

16.—SUNSPOTS AND SPRING RAINS AT GENEVA.

DIAGRAM : (*a*) Smoothed curve (5 av.) of rainfall in spring at Geneva from 1836. (*b*) Dotted curve, variation in number of rain-days in spring at Geneva. Continuous curve, smothed with averages of 5. (*c*) Sunspot curve (inverted).

The spring rains of Geneva since 1836 seem to have varied in a way suggestive of a relation to the sunspot cycle. Both raindays and rainfall exhibit the phenomenon, in which maxima of the smoothed curves are found coinciding with or closely following minima of the sunspot curve ; while the minima of the former have a similar relation to the maxima of the latter. (See *Nature*, Vol. 50, p. 475, and *Archives des Sciences*.)

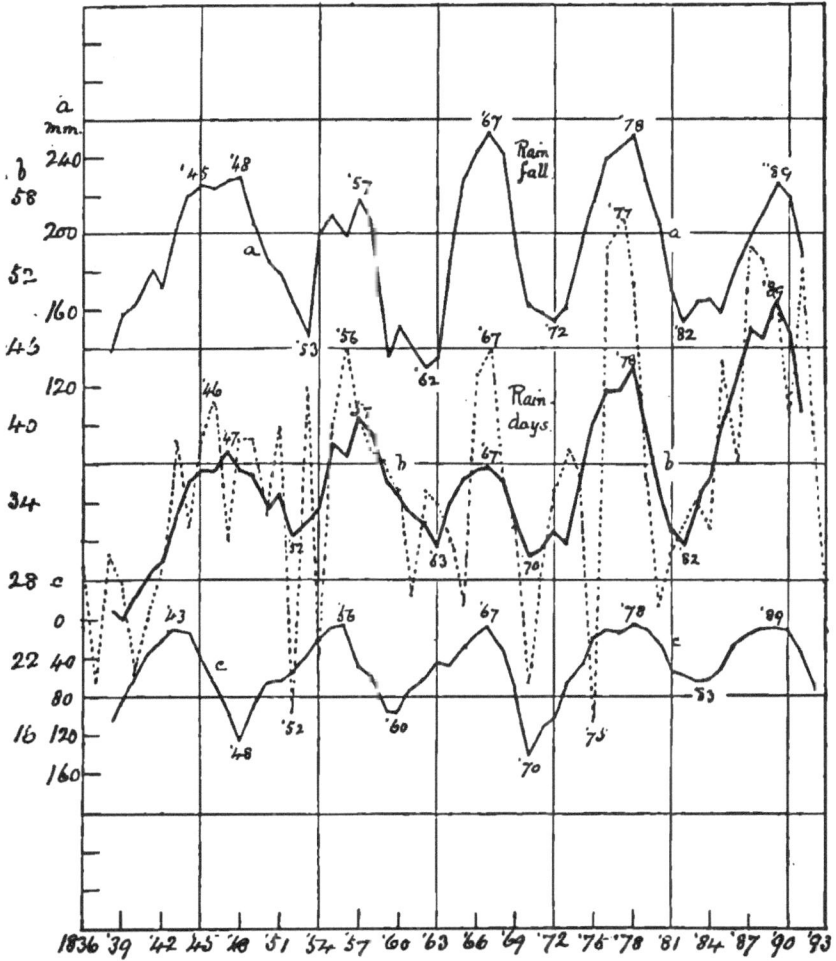

17.—SUNSPOTS AND THE MEAN TEMPERATURE OF THE FIRST QUARTER.

DIAGRAM : (*a*) Dotted curve, variation in mean temperature of the first quarter at Greenwich since 1841. Continuous curve, smoothed with averages of 5. (*b*) Sunspot curve (normal).

Our fourth and last example relates to the amount of cold in the first quarter of the year, the variations of this being measured by mean temperature.

The sunspot maxima being in
      1848   1860   1870 and 1883
we find maxima in the smoothed temperature curve in
      1850   1861   1870 and 1883 ;
mild first quarters, therefore, in general, accompanying or following many sunspots, and cold first quarters, few sunspots. The actual variations in this and other cases, are of course very zigzag in character ; and it will not be overlooked that we may find among the groups at the sunspot maxima first quarters with a low temperature, or among the groups at minima first quarters with a high temperature.

The oscillations of temperature here considered lie between 34·7° in 1855 and 44·2° in 1846 and 1872. (Buchan's Table, *loc. cit.*)

43

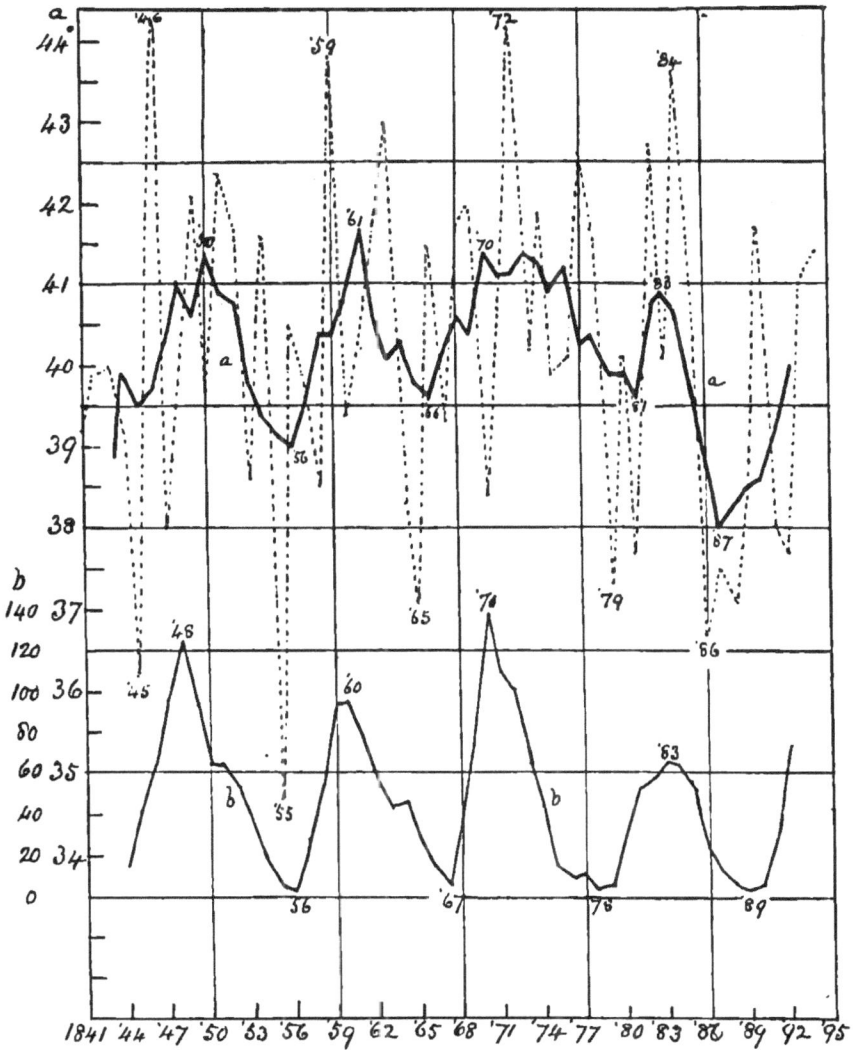

18.—EARLY AND LATE SEASONS. HARVEST DATES.

DIAGRAM : Dotted curve, variations in date (number of day in year) when wheat cutting was begun at Mere, in Wilts. Continuous curve, smoothed with averages of 10.

Some light on the general character of a season may be had from the dates at which various harvest operations commence in a locality. In a useful table Mr. T. H. Baker has given the dates when wheat cutting commenced in his district in Wiltshire, in a long series of years. The ordinary dates are each converted into the corresponding number of the day in the year. The upper parts of this dotted curve, then, it will be understood, represent late seasons (harvests), and the low points early ones.

These variations do not present much regularity. We might note that a very late harvest apparently comes after a lapse of about 19 years. Thus 1821 was very late ; then 1838 (17 years) ; then 1860 (22 years) ; then 1879 (19 years.) Again, we find very early harvests in 1826 ; in 1846 (20 years) ; in 1868 (22 years) ; and in 1893 (25 years) ; an average interval of 22 years. The smoothed curve descends to low points about 1829 and 1868, which are 39 years apart. It may be here mentioned that Brueckner supposes the weather of a great part of our globe to be subject to a cycle of about 35 years, cold and wet periods alternating with hot and dry ones ; and that 1830 and 1860 were approximately centres of hot and dry periods, while 1815, 1850, and 1880 were centres of cold and wet ones. Perhaps the smoothed curve may be said to harmonise so far with this view. (Baker's *Records of the Seasons*).

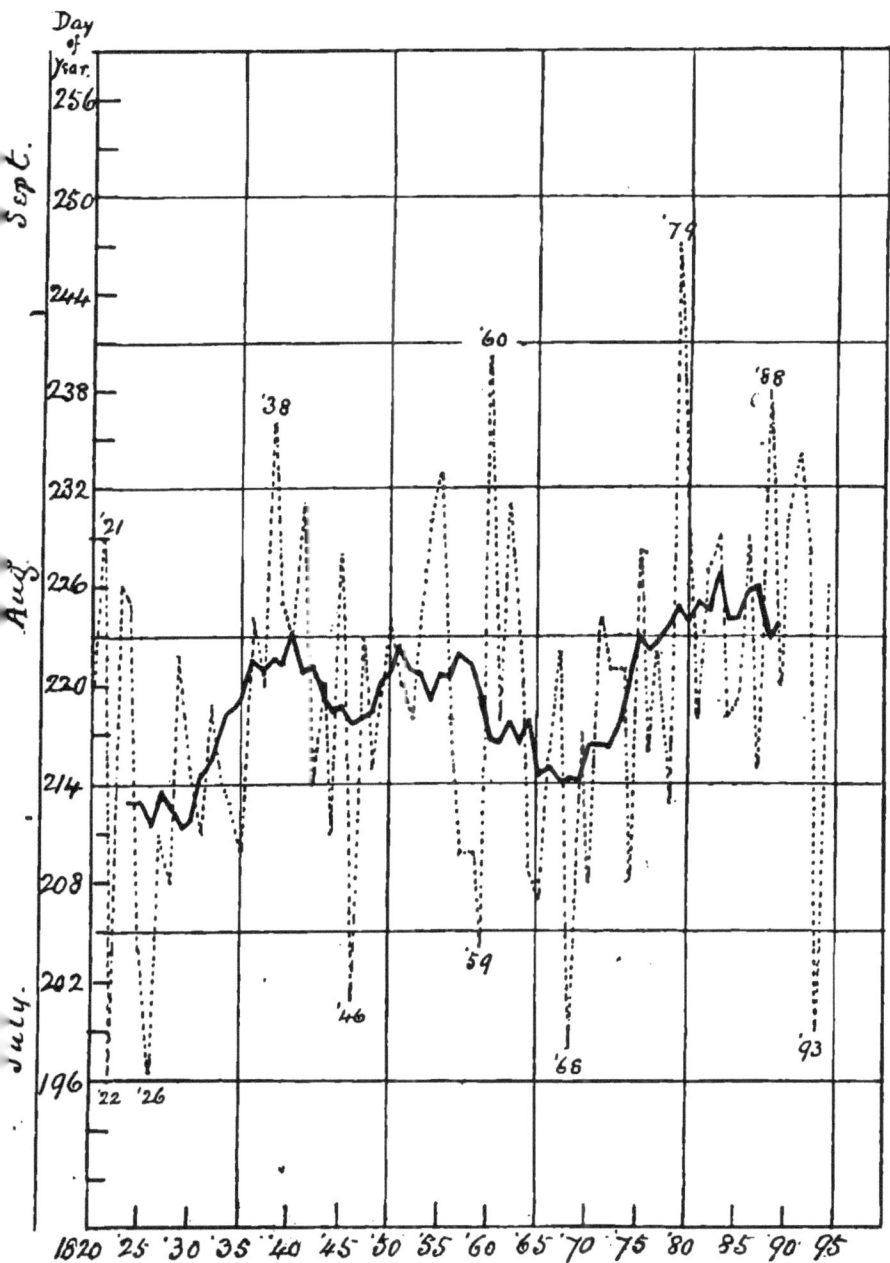

19.—EARLY AND LATE SEASONS. FIRST FLOWER-ING OF PLANTS.

DIAGRAM: (a–e) Five curves showing variations in date (day of year) of first flowering of (a) dog rose, (b) oxeye, (c) black thorn, (d) wood anemone, (e) coltsfoot, at Strathfield Turgiss, N.E. Hants. (f) Dotted curve, average of above. Continuous curve, smoothed with averages of 5. (g) Dotted curve, variations in date of first flowering of *Ribes Sanguineum* at Edinburgh (1850-1874). Continuous curve, smoothed with averages of 5.

We here consider early and late seasons from another point of view, that of the first flowering of plants, a branch of the science of phenology, (on which a valuable report is annually presented to the Royal Meteorological Society). The high points of those five curves (a–e) are again late dates, the low ones early dates. In the smoothed average curve below (f) we find a wave rising from a hollow about 1882 to a crest in 1887-1889. There was probably another crest about 1878. In the smoothed curve of *Ribes Sanguineum* at Edinburgh (g), to the left (with separate scale), we find two waves culminating in 1857 and 1866. It may not be amiss to note (without laying stress on it) that all those crests in the English and Scotch curves are near sunspot minima.

It should be borne in mind that all dates here considered are a great deal earlier than those of wheat cutting. It is right to state that those series for Strathfield Turgiss are not perfectly *homogeneous* throughout, but they cannot be far from the truth. (See Annual Reports on Phenology to the Royal Meteorological Society, by Rev. Mr. Preston and Mr. Mawley, in the *Quarterly Journal of the Royal Meteorological Society*.)

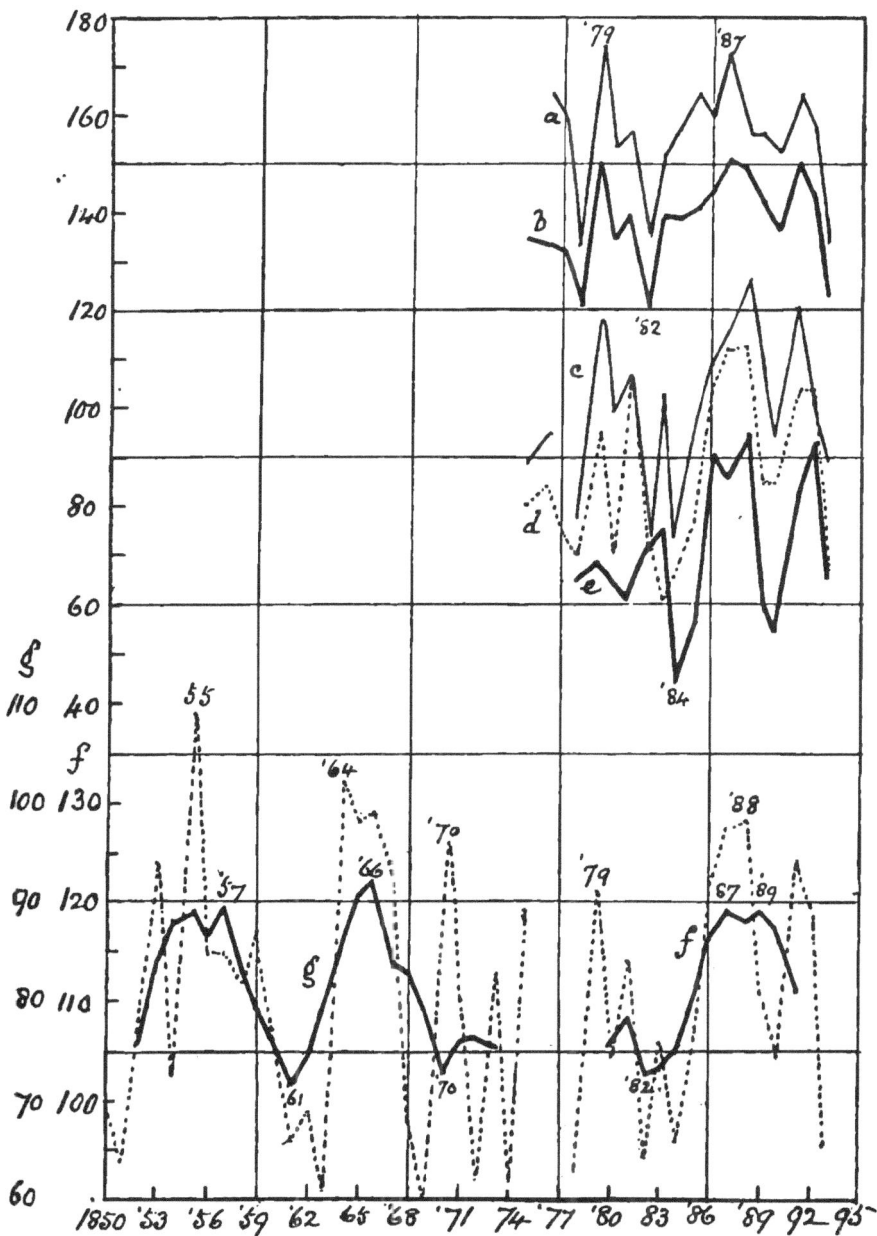

20.—LONDON FOG. EAST WINDS IN SPRING.

DIAGRAM : Upper curve, variation in number of days with fog in London in the winter half, since 1872. Lower curve, variation in number of days of easterly wind at Greenwich, since 1860.

LONDON FOG.—A definite and uniform system of recording fog is very desirable, and some uncertainty is at present felt in comparison of existing records, through want of agreement as to what shall be called a fog. The tables of London fog, prepared by Mr. F. J. Brodie, may, however, be taken as adequately showing how this unwelcome visitant has behaved in recent years. Confining our attention to the winter half (October–March), we find as conspicuously bad seasons 1873-74, 1879-80, 1886-87 (*i.e.* at 6 and 7 years' interval), the respective number of fog days being 59, 68, and 75. But the bad season 1890-91 rather breaks this regularity ; it had 69 fog days. Our winters of late have been much freer of fog.

EAST WINDS IN SPRING.—One of the annual tables issued by Greenwich Observatory gives, for each month of the year, the number of days of N. S. E. and W. winds (all directions reduced to these four). Adding the numbers for E. wind in the three spring months (March, April, May) since 1860, we get a series yielding the lower curve, from which it appears that the spring of 1880 was the most conspicuous for this character (38 days), while 1886 and 1893 come second and third (34 and 33). The earlier maxima seem to have been lower. (See Brodie's Paper on London Fog, *Quarterly Journal*, October, 1893, p. 233 ; *Annual Summary.*)

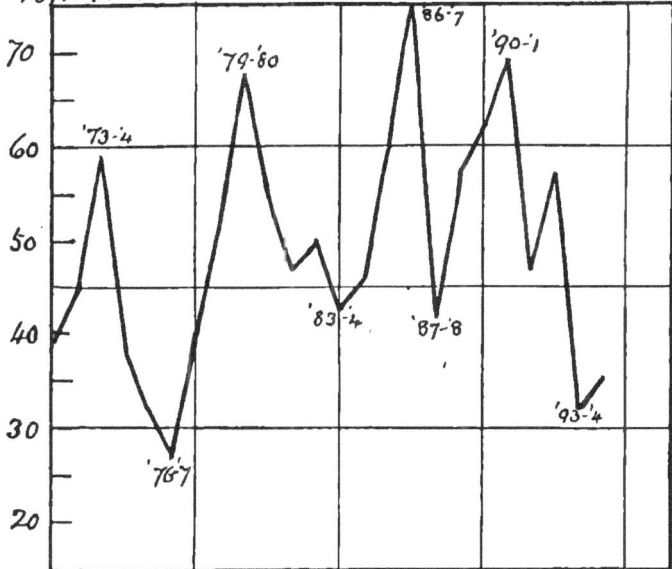

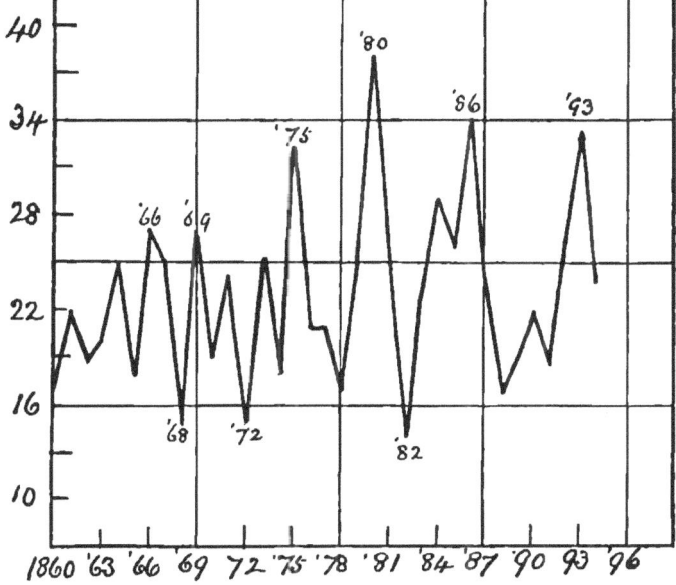

II.—DISEASE.

"Like crowded forest trees we stand,
And some are marked to fall."
Cowper.

I.—MEASLES.

DIAGRAM : (*a*) Dotted curve, variations of measles death rate in London since 1838. Continuous curve, smoothed with averages of 10. (*b*) Similarly smoothed curve, with separate scale, for England and Wales.

In the study of diseases we may, in a general way, discriminate between three groups. There are first the diseases which medical and sanitary science appears to be overcoming, (unless we are to interpret the recent decline in their fatality as merely the retiring slope of a long cyclical wave). Next, there are diseases which seem to be growing amongst us, spite of all applied knowledge (and the application of knowledge is of course by no means co-extensive with the knowledge itself). Lastly, there are diseases of which one can hardly affirm either permanent growth or decrease, they keep much about the same.

Our concern here will be chiefly with diseases of the zymotic class, and we may begin with measles.

The smoothed London curve shows several long waves, with crests about 1843 (?), 1862, and 1885 (*i.e.*, at 19 and 23 years' interval), and hollows about 1852 and 1871 (19 years). But on the whole we seem to remain *in statu quo*. The epidemics of 1839 and 1845, indeed, appear to have been much more severe than any since, reaching the figures 1,133 and 1,122 per million living. The rate goes down on the other hand to 246 in 1852. The smoothed curve for England and Wales resembles that for London.

Measles, in this country, has two maxima in the year ; the larger in November–January, the smaller in May–June. It must be remembered that here and elsewhere we deal only with mortality, and the prior time of incubation and illness should be taken into account in considering seasonal influences. It is probable that measles is due to some microbe, but this has not yet been identified. (*Annual Summary ; Annual Report of Registrar General.*)

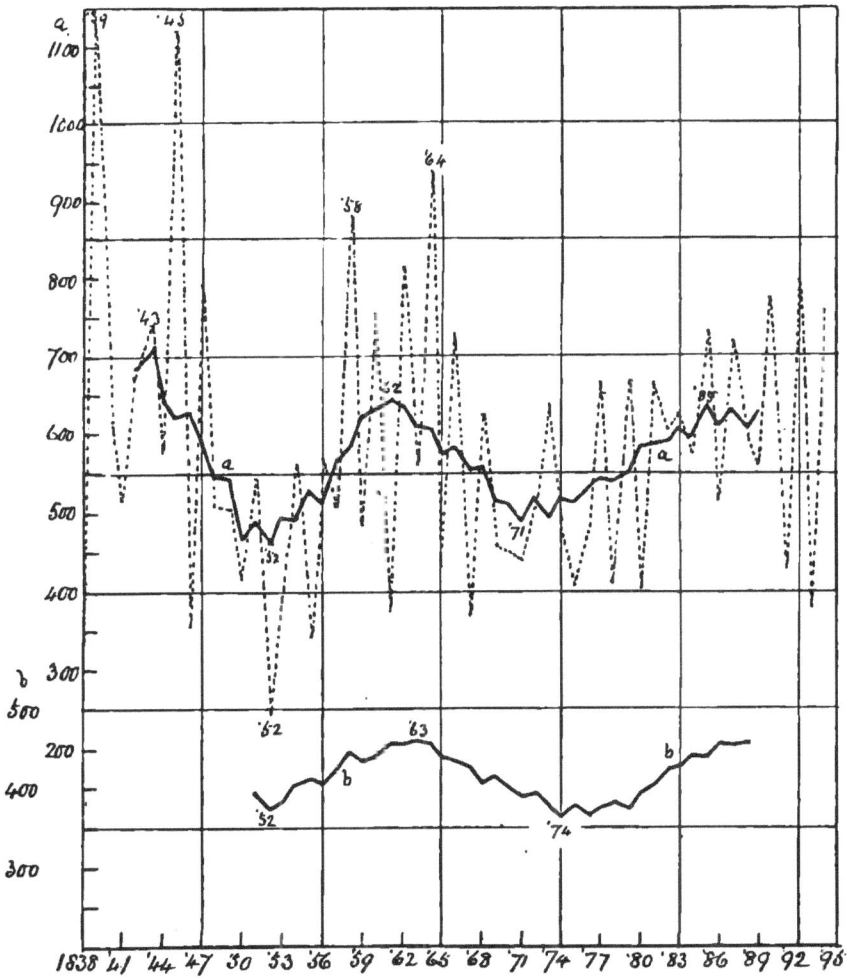

2—DIPHTHERIA.

DIAGRAM : (*a*) Dotted curve, variations of diphtheria death rate in London since 1859. Continuous curve, smoothed with averages of 10. (*b*) Similarly smoothed curve, with separate scale, for England and Wales.

The increase of diphtheria in recent years is notorious. Our curves begin at 1859, when the disease began to be registered separately (from scarlet fever). The London death rate, which was rather high in some of the fifties and sixties (284 in 1859, 275 in 1863), went down to a minimum in 1872 (80) ; since which year it has been rising on the whole, till in 1893 the enormous rate of 760 was reached. In 1894 there is some improvement. The rise in England and Wales has been less rapid.

Some allowance is probably to be made for improved diagnosis and registration ; but even so, a large increase (in London) is unmistakeable.

The alarming spread of diphtheria in recent years has been attributed to the growing aggregation of children in schools, &c. ; but there is some reason to think the disease is subject to a long cycle, a wave crest of which we have been lately approaching.

Diphtheria is far more common in temperate and cold regions than in the tropics. In this country it is most common in the last quarter, and lowest in the summer months. Cold and damp apparently increase the susceptibility to infection.

The microbe of diphtheria is that known as Löffler's *bacillus.* (*Annual Summary ; Annual Report of Registrar General.*)

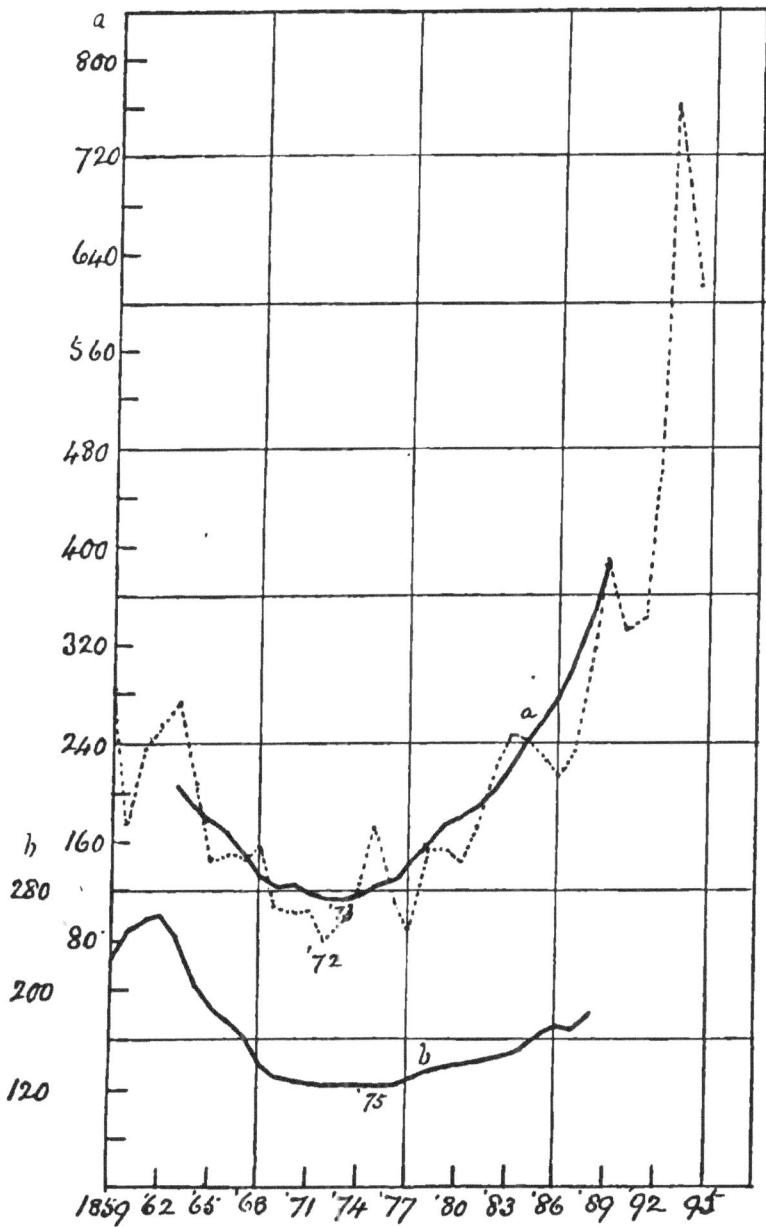

3—WHOOPING COUGH.

DIAGRAM : Method as in 1 and 2.

This widespread disease, so fatal to children, appears to have been declining, on the whole, in recent years (since the sixties). The 10 years' average for London in 1857 was 921 (per million living) ; but in 1889 it had got down to 628.

The London rate was very high in 1861 (1,260), also in 1841, 1869, 1878, and 1882, (all over 1,180). Its lowest point was reached in 1883 (410). A relatively high point is reached, it will be seen, every 3 to 5 years.

In London, whooping cough increases steadily from its lowest in September to a maximum in April, from which it declines through the warmer months. It is often observed to become worse when the wind turns cold or suddenly changes. Its specific microbe (if it have one) is not yet identified. (*Annual Summary; Annual Report of Registrar General.*)

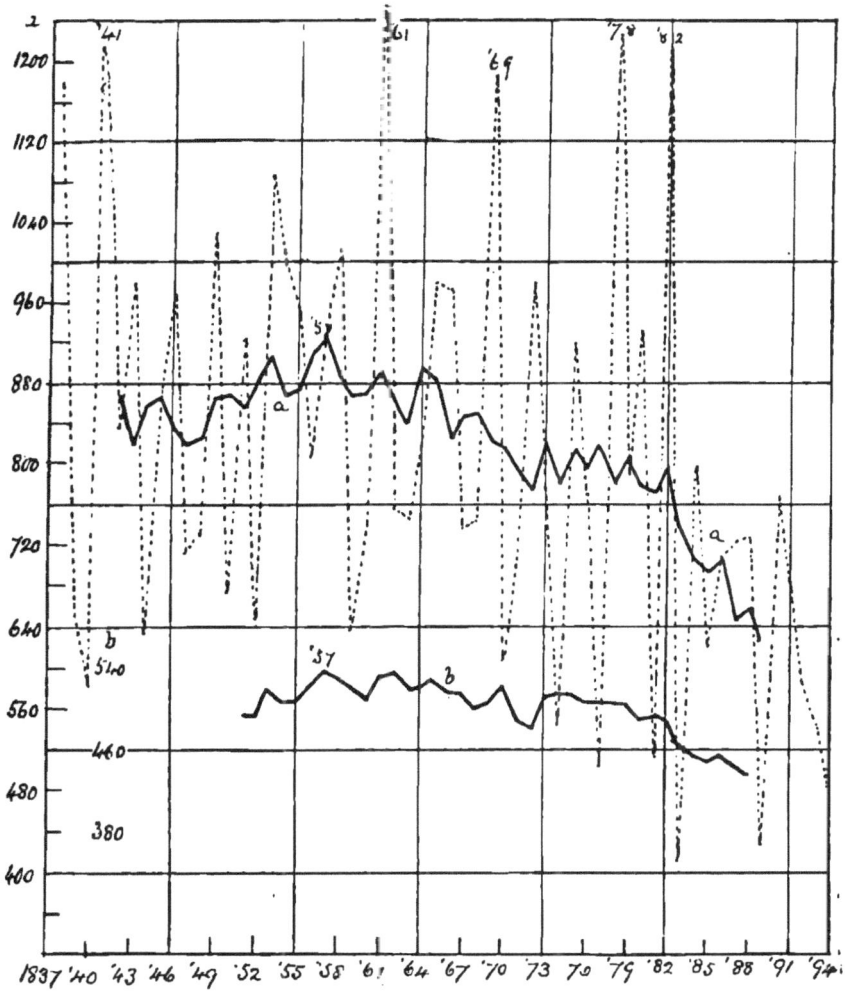

4.—SCARLET FEVER.

DIAGRAM : Method as in 1 and 2.

This disease presents a still more evident decline in recent years, (since the sixties). From a 10 years' average of 1,133 in 1865, the smoothed London curve descends to one of 240 in 1889. The extreme variations of the actual death rate in London are 2,131 in 1848, and 142 in 1891.

The variations in the death rate of this disease have been pretty regular in their recurrence. The dotted curve reaches wave crests every 4–7 years. The prevalence and mortality of scarlet fever are greatest in the autumn and least in the spring, and its seasonal curve is thus, roughly, opposite to that of whooping cough. The maximum is in October. Curiously, in New York, scarlet fever is most prevalent in spring and least in autumn.

According to Dr. Ballard, a temperature above the average for the season, and a dry atmosphere, with little rain, favours the prevalence of scarlet fever more than the reverse conditions. The disease has been connected with a certain *streptococcus* as its producing cause.

One is disposed to ask whether the long wave of temperature indicated in I. 8 may not have something to do with the course of this and some other diseases (whooping cough, diphtheria)? (*Annual Summary ; Annual Report of Registrar General.*)

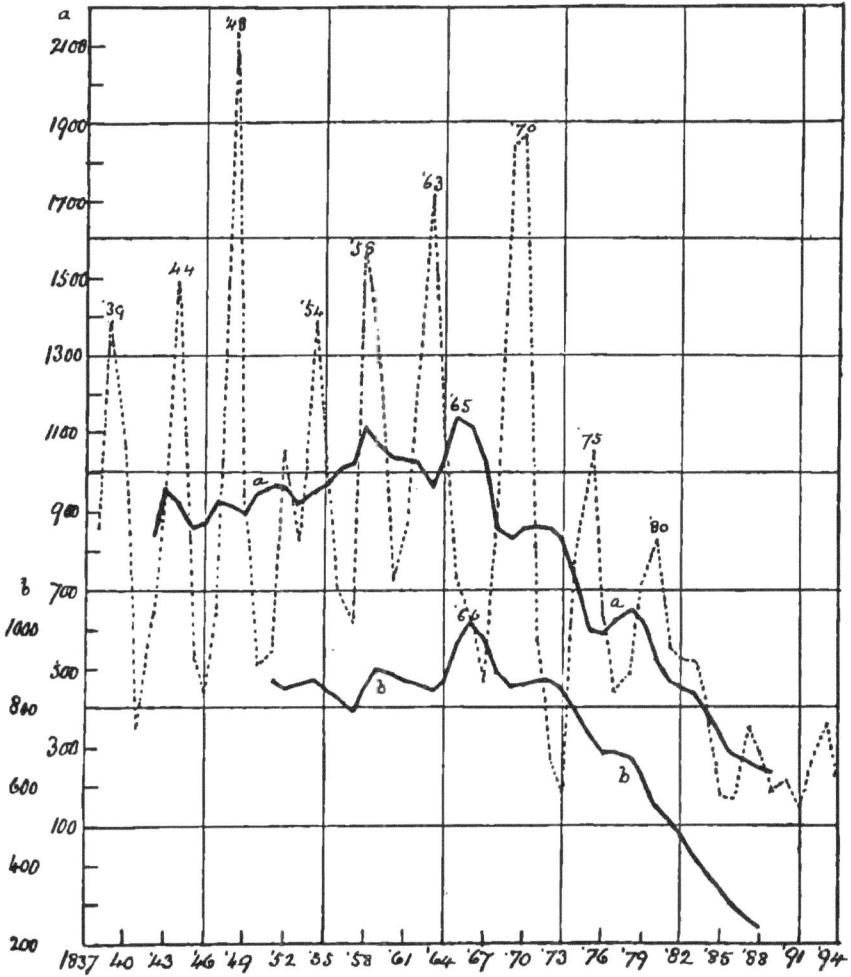

5.—ENTERIC OR TYPHOID FEVER.

DIAGRAM : Method as in 1 and 2.

Our curves here extend only from 1869, the time when enteric fever began to be registered separately from typhus and simple or ill defined fever. Note the steady decline in both curves ; from a death rate of 337 in 1869, the London curve goes down to 102 in 1892.

The prevalence and mortality of this disease are greatest in the autumn and least in the spring and early summer. The *bacilli* known as Gaffky's are believed by many pathologists to be the cause of enteric fever.

This fever, it is well known, is distinguished from typhus fever by the intestinal lesions present. The mortality from typhus fever has also greatly declined of late years. (*Annual Summary ; Annual Report of Registrar General.*)

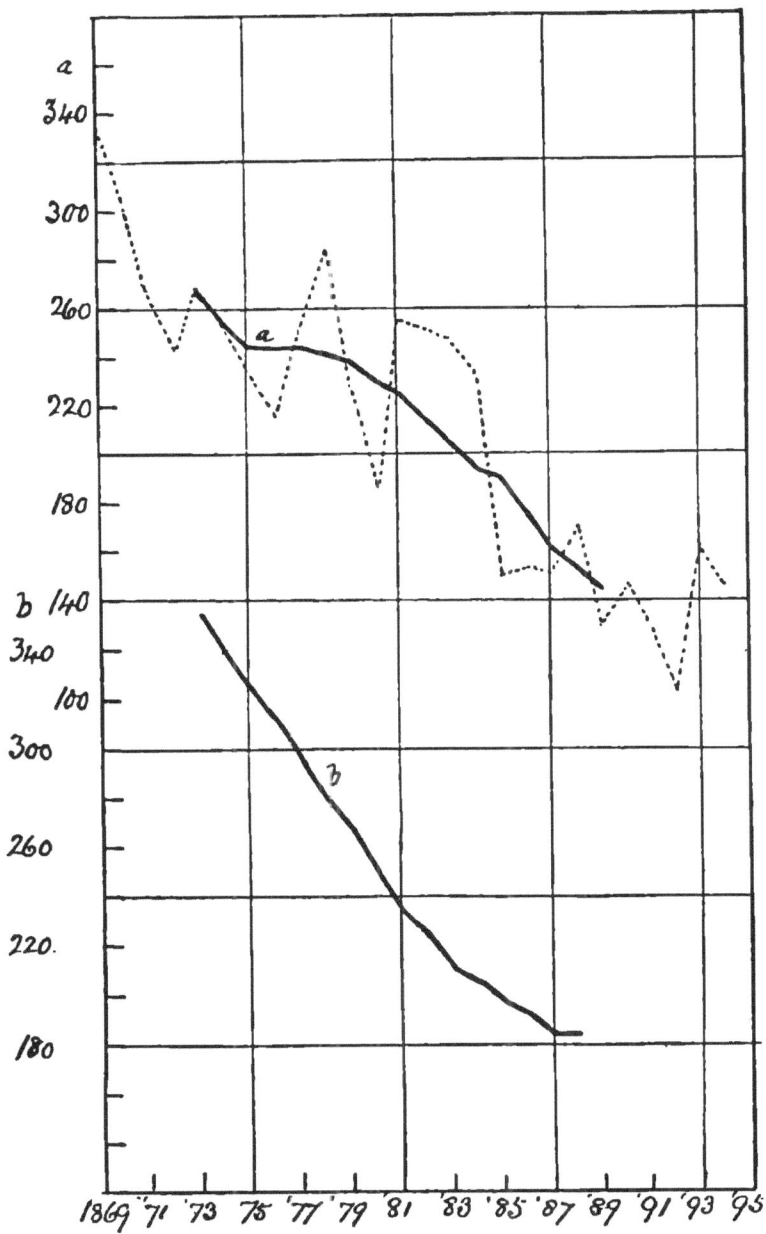

6.—DIARRHŒA AND DYSENTERY.

DIAGRAM : (*a*) Dotted curve, variations of the death rate in London since 1838. Continuous curve, smoothed with averages of 5. (*b*) Similarly smoothed curve for England and Wales. (*c*) Similarly smoothed curve of the mean temperature of July at Greenwich.

In the London smoothed curve we may notice three long waves culminating in 1851, 1870, and 1885, (19 and 15 years apart). The range of the death rate is from 1,683 in 1849 to 253 in 1839. Recent values are still above those at the beginning of the curve.

Diarrhœa is now reckoned as largely of epidemic character. With regard to its association with heat, it would appear from recent researches that atmospheric temperature is of less importance in relation to it than earth temperature. According to Dr. Ballard, the summer rise of diarrhœal mortality does not commence till the mean temperature of a 4 ft. deep earth thermometer reaches about 56°, no matter what temperature has previously been reached by the atmosphere. The two (death rate and earth temperature) reach their maximum in the same week, and the decline of both is similarly slow. The average summer wave of diarrhœa in London rises to a maximum about the end of July : it is mainly due to infantile diarrhœa. We may trace a considerable amount of correspondence, (from the nature of the case it could not be exact), between the smoothed July temperature curve (*c*) and the London diarrhœa curves (*a*). Calm weather promotes diarrhœa mortality, while wind lessens it. Rainfall affects it indirectly through the temperature of the ground. Moderate dampness of soil favours the disease, but excessive wetness or dryness are thought to lessen it.

Several kinds of *bacilli* have been connected with the disease. (*Annual Summary ; Annual Report of Registrar General*, &c.)

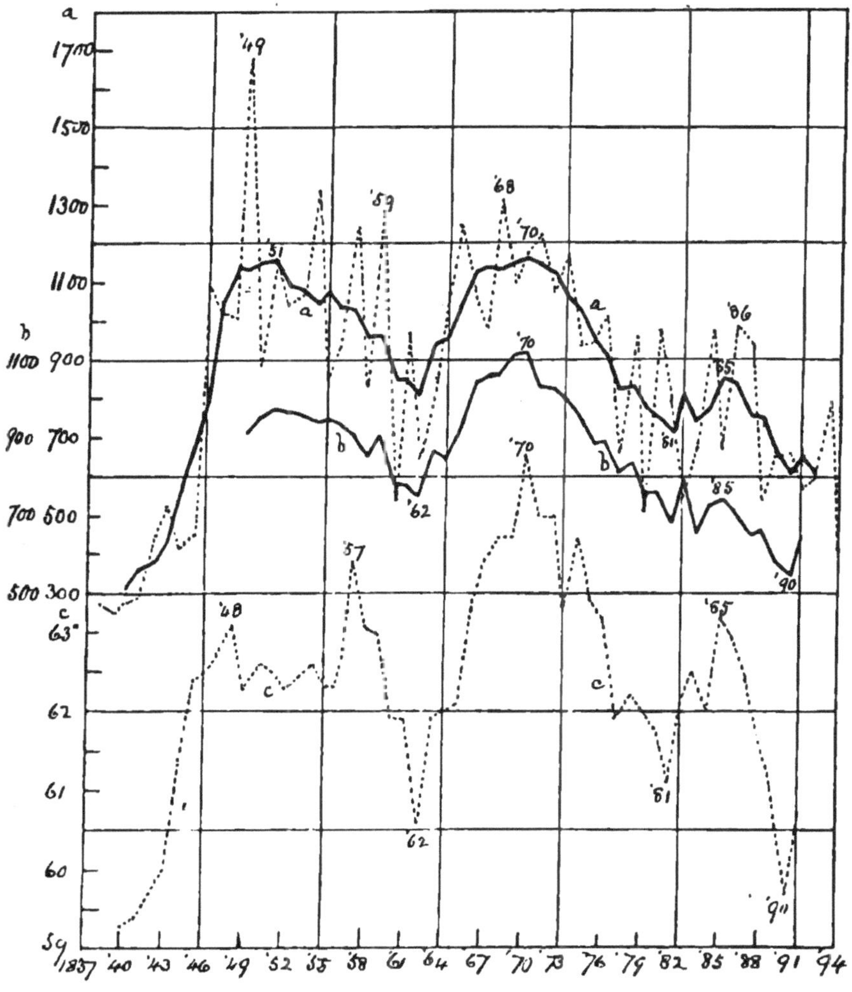
a
1700
1500
1300
1100
b
1100 900
900 700
700 500
500 300
c
63°
62
61
60
59
1837 '40 '43 '46 '49 '52 '55 '58 '61 '64 67 '70 '73 '76 '79 '82 '85 '88 '91 '94

7.—SMALL POX SINCE 1838.

DIAGRAM : (*a*) Actual variations of the small pox death rate in London. (*b*) Ditto, in England and Wales (data lacking 1843–1846).

Excluding the great epidemics of 1838 and 1871–72 a gradual decline is apparent, especially in the lower curve. In the years 1838 and 1871 the London death rate rose to 2,168 and 2,422 respectively (per million living). In 1889 the return is *nil*. In the next 4 years there is a gradual rise (represented by the figures 1, 2, 10, 48), checked in 1894, however. In the London curve relative maxima are reached every 3–6 years.

In Oriental countries the spread of small pox appears to be favoured by cold and retarded by heat. In Europe the mortality is greatest in spring and autumn. (*Annual Summary; Annual Report of Registrar General.*)

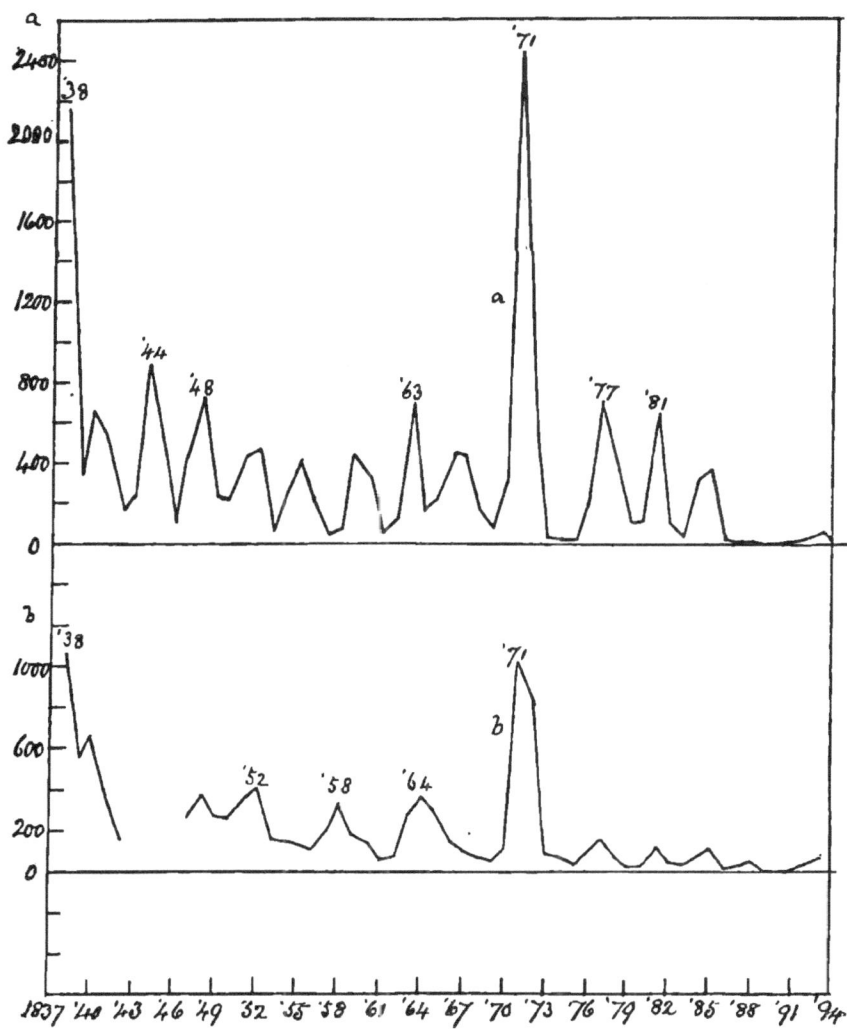

8.—TWO CENTURIES OF SMALL POX.

DIAGRAM : Ratio of deaths from small pox to 1,000 deaths from all causes in London, averaged in decades (1701-10, &c.).

In a valuable paper contributed to the Statistical Society, Dr. Guy has given statistics of 250 years of small pox in London, showing, year by year, the ratio of deaths from small pox to 1,000 deaths from all causes. Extending the table to 1891 we have made decadal averages of those ratios since 1701. The data between 1830 and 1840 are defective.

The curve rises high in the latter half of last century (average for 1861-1870, 102·5), but in this century there has been a rapid drop, and in the last decade we come to an average of 6·9.

With regard to the vexed question of vaccination, some dates may here be given. The practice was introduced by Jenner about the end of last century (1798). Vaccination was optional before 1853 ; then from 1854-1871 it was obligatory, but not efficiently enforced ; and since 1871 it has been more efficiently enforced. (See Dr. Guy's Paper to Royal Statistical Society, Vol. XLV.)

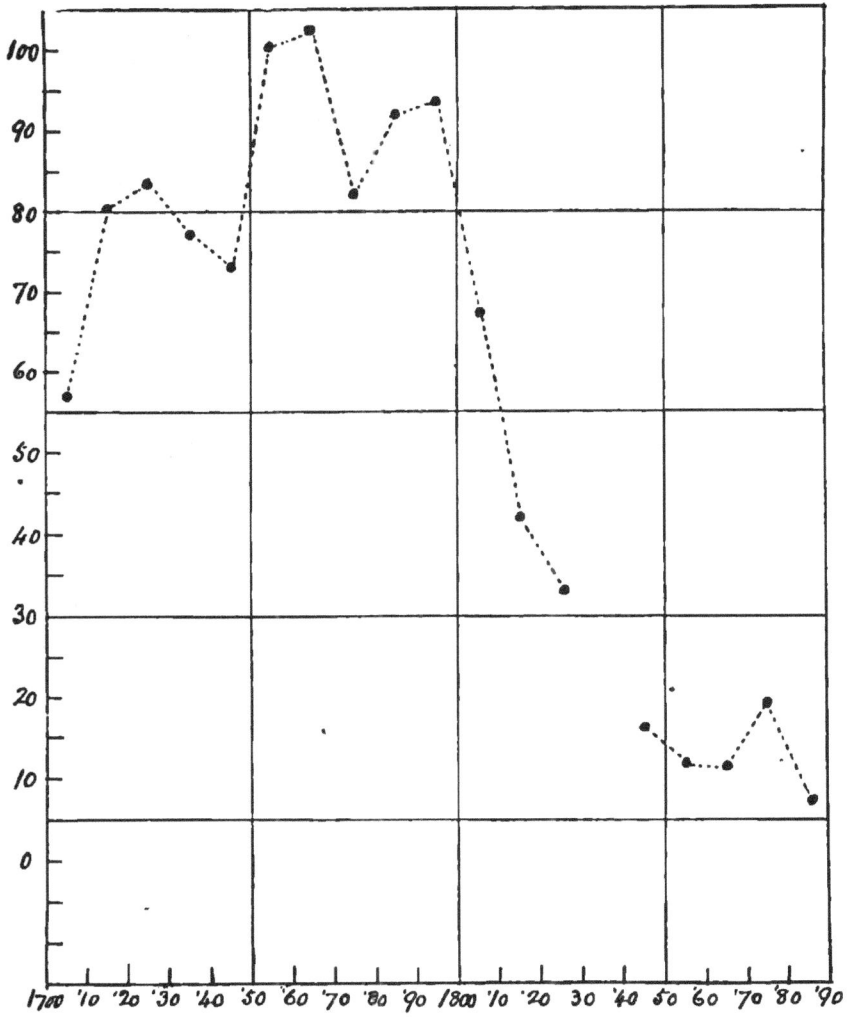

9.—PROGRESSION IN THE SUMMER DEATH RATE WAVE.

DIAGRAM : (*a*) Dotted curve, weekly death rate, Plymouth, from 22nd to 39th week of 1893. Continuous curve, smoothed with averages of 5. (*b*) Ditto for Cardiff. (*c*) Portsmouth. (*d*) Brighton. (*e*) Hull.

There is generally a well-marked wave in the summer death rate, and if we compare these waves for different places, some interesting relations are met with. Here we take the hot summer of 1893, and show the death rate variations from the 22nd to the 39th week (*i.e.*, say, June to September) at five places. The smoothed curves bring out more plainly the retarded culmination of these waves in an easterly direction (which is probably a temperature effect). Thus we have the Plymouth wave crest about the 25th week ; Cardiff, the 26th ; Portsmouth, the 29th ; Brighton, the 34th. In the north, Hull is still later, 35th week. This kind of relation may be observed in other years, but is not invariable. (*Annual Summary.*)

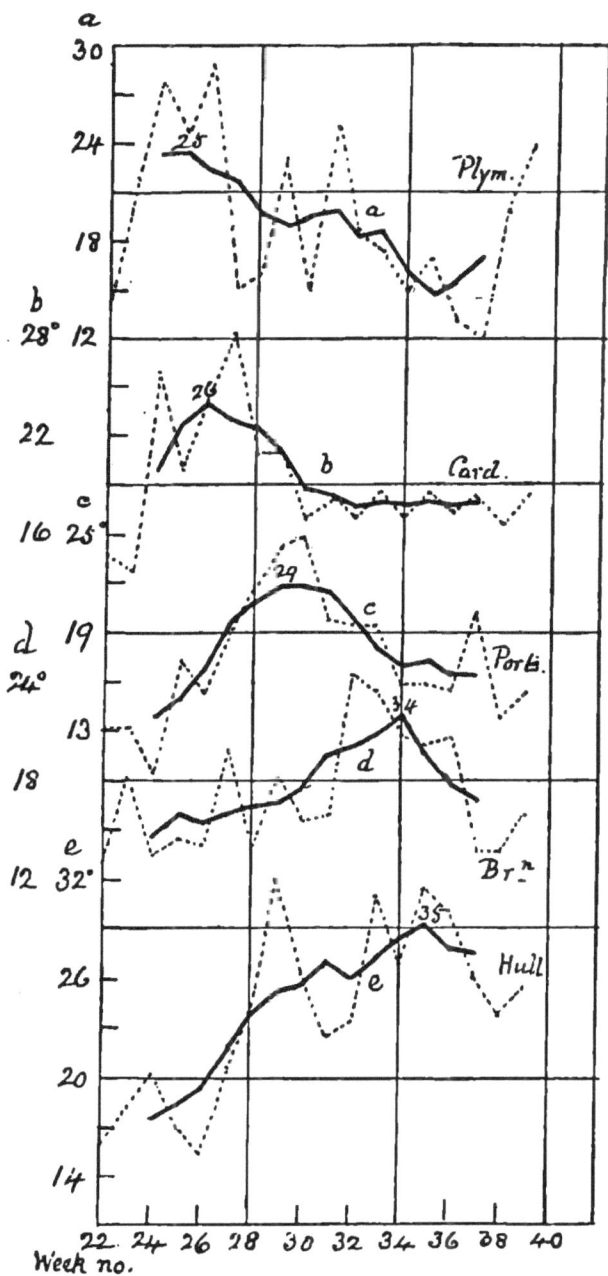

a

30
24
18
b
28° 12

22

16
c
25°
19
d
24°
13

18
e
12 32°

26

20

14

22 24 26 28 30 32 34 36 38 40
Week no.

Plym.
Card.
Ports.
Br.ᵃ
Hull

10.—WAVES OF INFLUENZA.

DIAGRAM : Weekly deaths in London in the five epidemics.

The five recent successive visitations of influenza are here depicted. The worst of the series, as regards highest mortality in a week, was that in the winter of 1891-92 ; there were as many as 506 deaths in the week ending January 23rd. The last epidemic (1895) comes very near this however, with 473 deaths in the week ending 9th March. Both of these waves are sharper in their culmination than the wave of 1891.

We may note the dates of maxima in those 5 curves : week ending January 18th, May 23rd, January 23rd, December 16th, March 9th.

Hirsch considers that influenza is independent of seasons, and of the influences of weather.

The last previous group of influenza epidemics was about the forties ; 1847 and 1848 were very bad years. The chief remaining epidemics of the century were in 1803, 1833, and 1837-38. The disease apparently alters in type from time to time. There is reason to believe that a particular *bacillus* is in causal relation to it. (*Weekly Return.*)

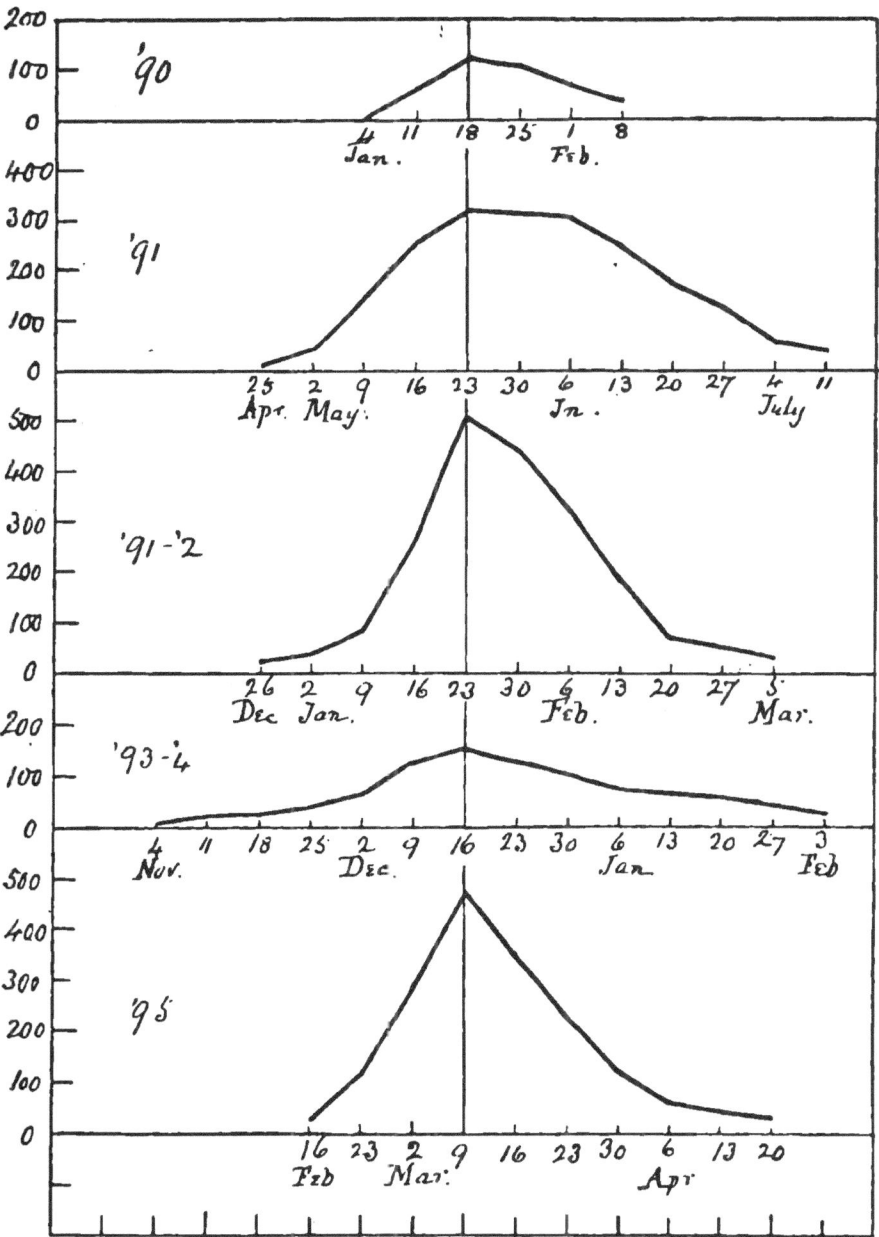

'90

200 — 100 — 0

4 11 18 25 1 8
Jan.   Feb.

'91

400 — 300 — 200 — 100 — 0

25 2 9 16 23 30 6 13 20 27 4 11
Apr. May.   Jn.   July

'91-'2

500 — 400 — 300 — 200 — 100 — 0

26 2 9 16 23 30 6 13 20 27 5
Dec Jan.   Feb.   Mar.

'93-'4

200 — 100 — 0

4 11 18 25 2 9 16 23 30 6 13 20 27 3
Nov.   Dec.   Jan   Feb

'95

500 — 400 — 300 — 200 — 100 — 0

16 23 2 9 16 23 30 6 13 20
Feb.   Mar.   Apr

11.—QUARTERLY DEATHS FROM WHOOPING COUGH AND MEASLES IN LONDON.

DIAGRAM : (*a*) Whooping cough. (*b*) Measles. (1.'76 means first quarter of 1876, &c.)

It is instructive to put into curve form the numbers of deaths from a given disease, *in quarters*, in a series of years. The diagram gives those for whooping cough and measles in London since 1875. Each space between two marks at the foot indicates a year, and is divided into four. The whooping cough curve, it will be seen, reaches a maximum with great regularity about every two years ; the first quarter of 1876, the second of 1878, the first of 1880, and so on. The worst quarter was the first of 1882, when there were 2,224 deaths. The measles curve also reaches maxima at about two years' intervals, and these maxima evidently tend to increase in height of late : the highest figure is in the second quarter of 1894 (1,745 deaths).

Comparing the maxima of the two curves, it will be noticed that through a great part of the period they alternate ; measles one year, whooping cough the next, and so on. (*Quarterly Reports of Registrar General.*)

73

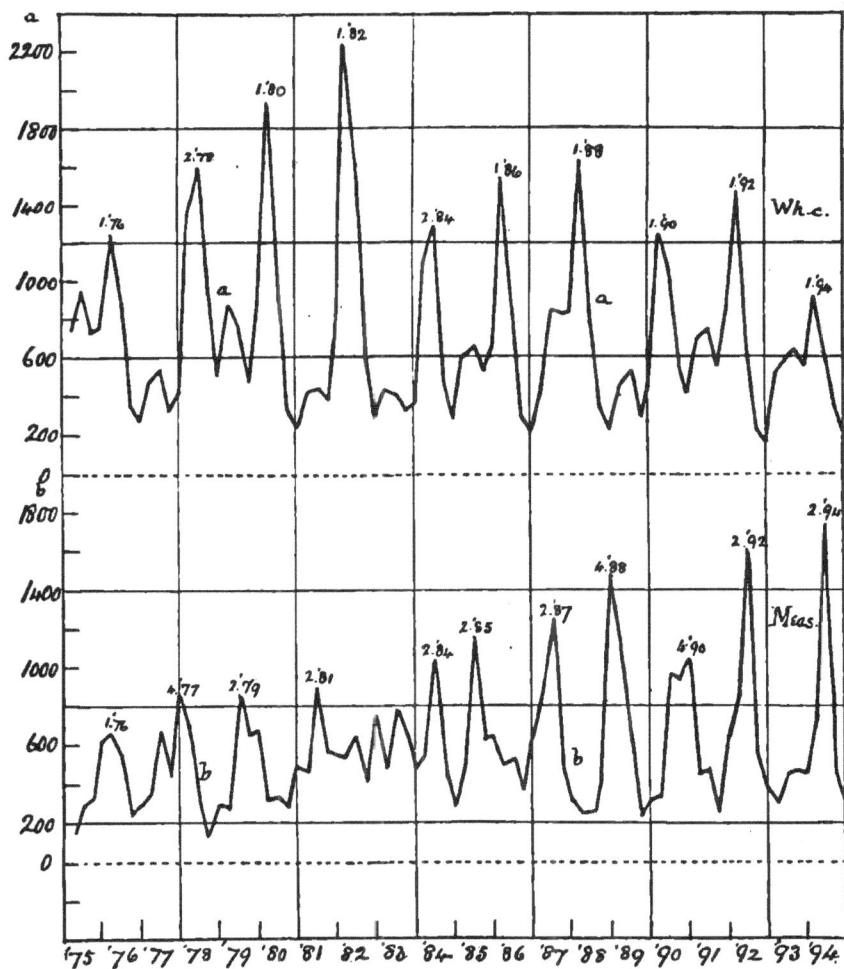

12.—THE INCIDENCE OF DISEASE IN DIFFERENT TOWNS.

DIAGRAM : (*a*) Quarterly deaths from whooping cough in Bristol since 1875. (*b*) Ditto, in Portsmouth. (*c*) Quarterly deaths from scarlet fever in Bristol. (*d*) Ditto, in Leeds.

Our large towns, even when not far apart, may be quite differently affected by disease. In illustration of this, two comparisons are here given. In the case of whooping cough at Bristol and Portsmouth, these quarterly curves show (during much of the period considered) that when the Bristol curve has culminated, the Portsmouth curve has been low ; and when the latter has culminated, the Bristol curve has been low. This is not, however, an unvarying rule. Note, further, that the intervals between crests in both cases tend to be a little wider than in London.

The other case is that of scarlet fever in Bristol and Leeds. While the Bristol curve has maxima in 1875, 1880, and 1887, Leeds has them in 1875, 1878, and 1883. (*Quarterly Reports of Registrar General.*)

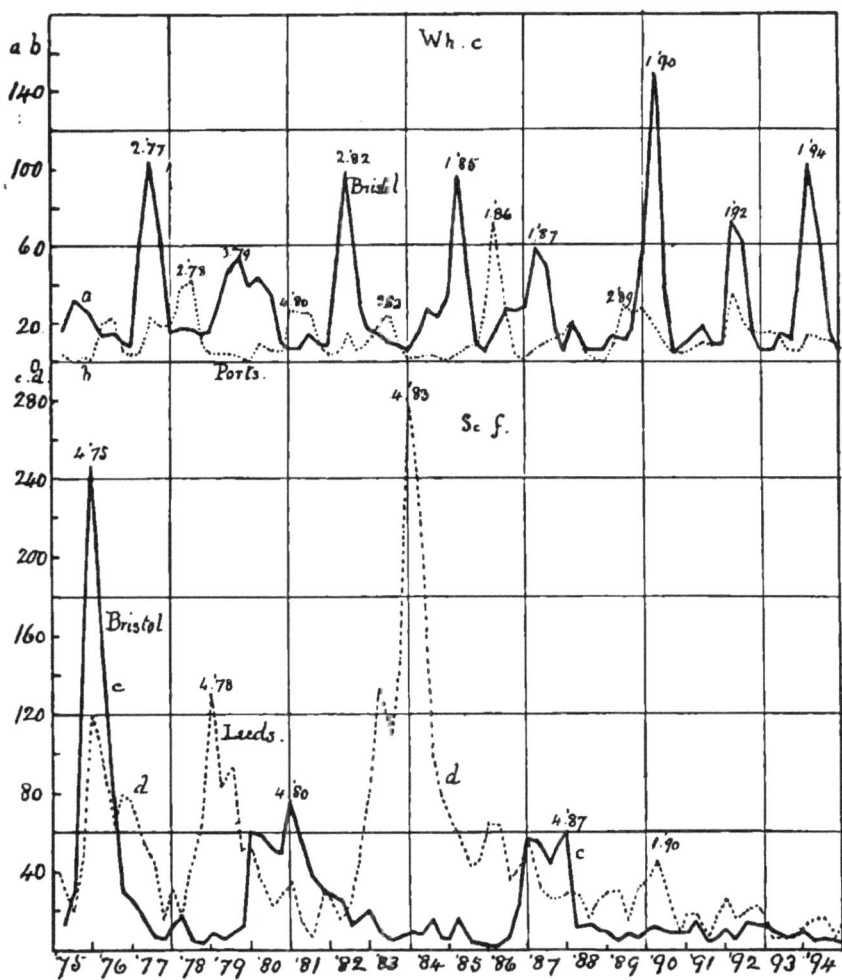

13.—A GENERAL VIEW OF DISEASE IN LONDON IN 1893–94.

DIAGRAM : (*a*) Smoothed curve (5 av.) of weekly deaths from diseases of the respiratory system in London. (*b*) Ditto, from the chief zymotic diseases. (*c*) Ditto, from phthisis. (*d*) Ditto, from diseases of the circulatory system.

The years 1893 and 1894 were very opposite with regard to weather, which may be considered an advantage from our present standpoint.

In this group of smoothed curves we may first note the preponderance of diseases of the respiratory system, and the high point reached by this curve in winter, (719 in the 49th·week of 1893, *i.e.*, early in December). In both descents of the curve from the maxima there is a curious reversal. Next we have the curve of zymotic diseases (including small pox, measles, scarlet fever, diphtheria, whooping cough, typhus, enteric fever, simple continued fever, diarrhœa and dysentery, and cholera), with its striking summer wave, (diarrhœa chiefly), in the very hot summer of 1893 ; in the 27th week the average reached 461. In 1894 we find a different state of things, two much smaller waves culminating in the 19th (310) and 31st (303) weeks respectively, the first due to measles chiefly, the second to diarrhœa.

Lastly, we have curves for phthisis and for diseases of the circulatory system, both culminating in weeks of winter, (phthisis, 186 deaths in the 52nd week of 1893, circulatory system 160 in the 49th week of 1893). This classification of disease is of course not exhaustive. (*Weekly Return.*)

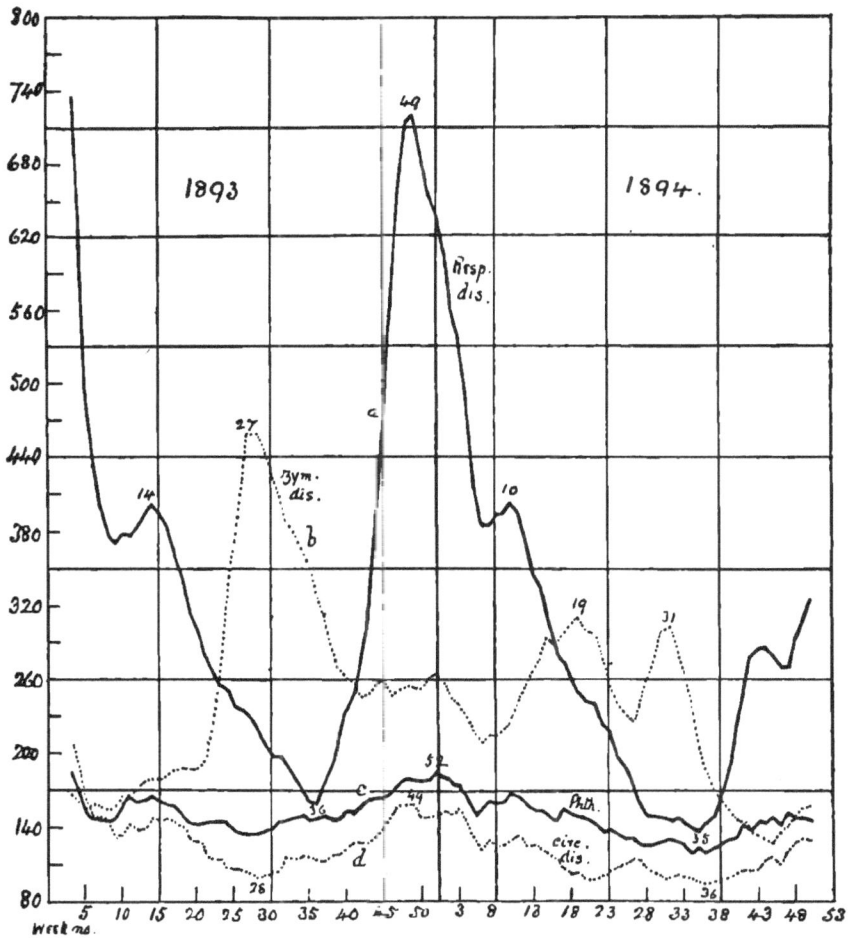

14.—SOME ZYMOTIC DISEASES IN LONDON IN
1893–94.

DIAGRAM : (*a*) Smoothed curve (5 av.) of weekly deaths from measles in London. (*b*) Ditto, from diphtheria. (*c*) Ditto, from whooping cough. (*d*) Ditto (with separate scale), from scarlet fever.

We have just looked at zymotic diseases as a whole ; we may now consider some members of the group in the same way.

Here we at once notice the high measles wave of 1894, going up to 162 in the 21st week, a great contrast to that of the previous year. Diphtheria reaches a pronounced maximum in the 45th week of 1893, and whooping cough in the beginning of 1894.

The scarlet fever curve (given with separate scale below to avoid confusion) also culminates in the last quarter of 1893, and in general the values for that year are above those for the second year. (*Weekly Return.*)

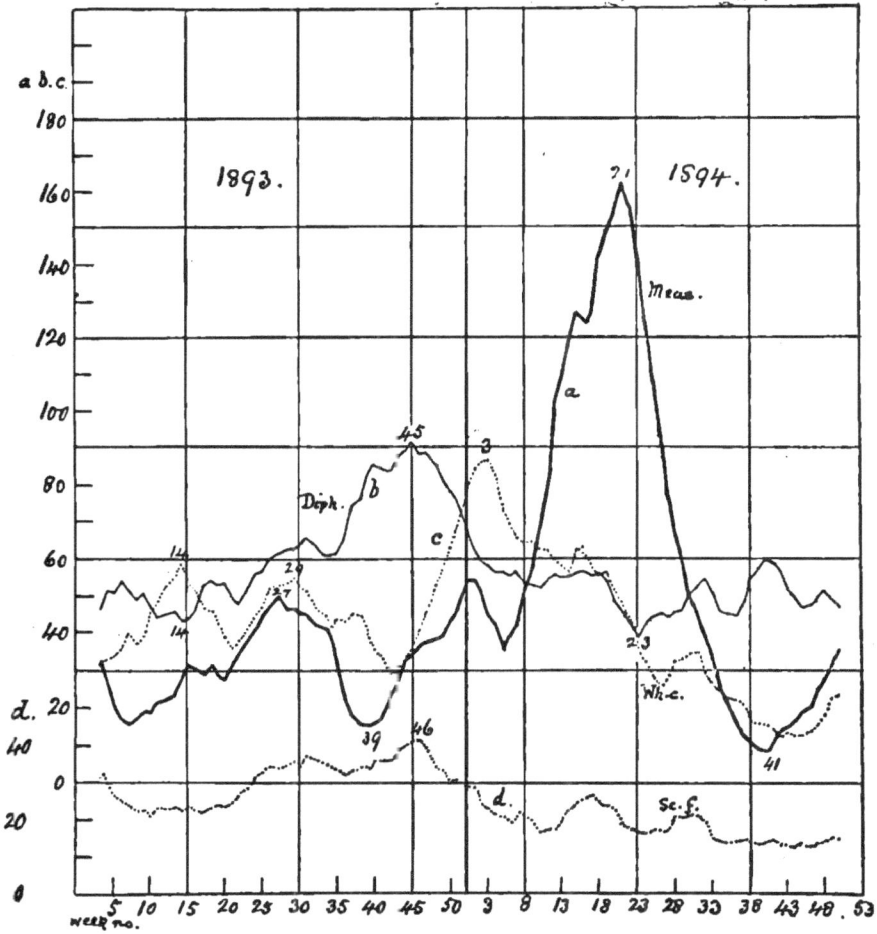

15.—RESPIRATORY DISEASE IN LONDON IN THE FIRST QUARTER.

DIAGRAM : (a) Dotted curve, variation in first quarter death rate (per 1,000) from diseases of the respiratory organs, in London, since 1855 (*inverted*). Continuous curve, smoothed with averages of 5. (b) Smoothed curve (5 av.) of mean temperature in first quarter at Greenwich.

This smoothed and inverted curve of respiratory disease in the first quarter exhibits two well-marked waves with crests (low death rate) in 1870 and 1883.

Respiratory disease is of course closely associated with cold weather. A spell of cold weather, in winter, is generally followed by a rise in the deaths from bronchitis, &c. We have already considered the mean temperature of the first quarter (No. 17 in first part), and we repeat the smoothed curve of it since 1755. There is a fair amount of correspondence with the respiratory curve (from the nature of the case we could not expect it to be very exact), the temperature wave showing maxima in 1870 and 1883, and the average death rate being then low. Now the temperature curve we found to correspond fairly with the sunspot curve. But to say that we cough more or less, according as the sun's face is less or more spotted, might argue some fitness for the attention of Commissioners in Lunacy. (*Annual Summary;* Buchan's Table, *loc. cit.*)

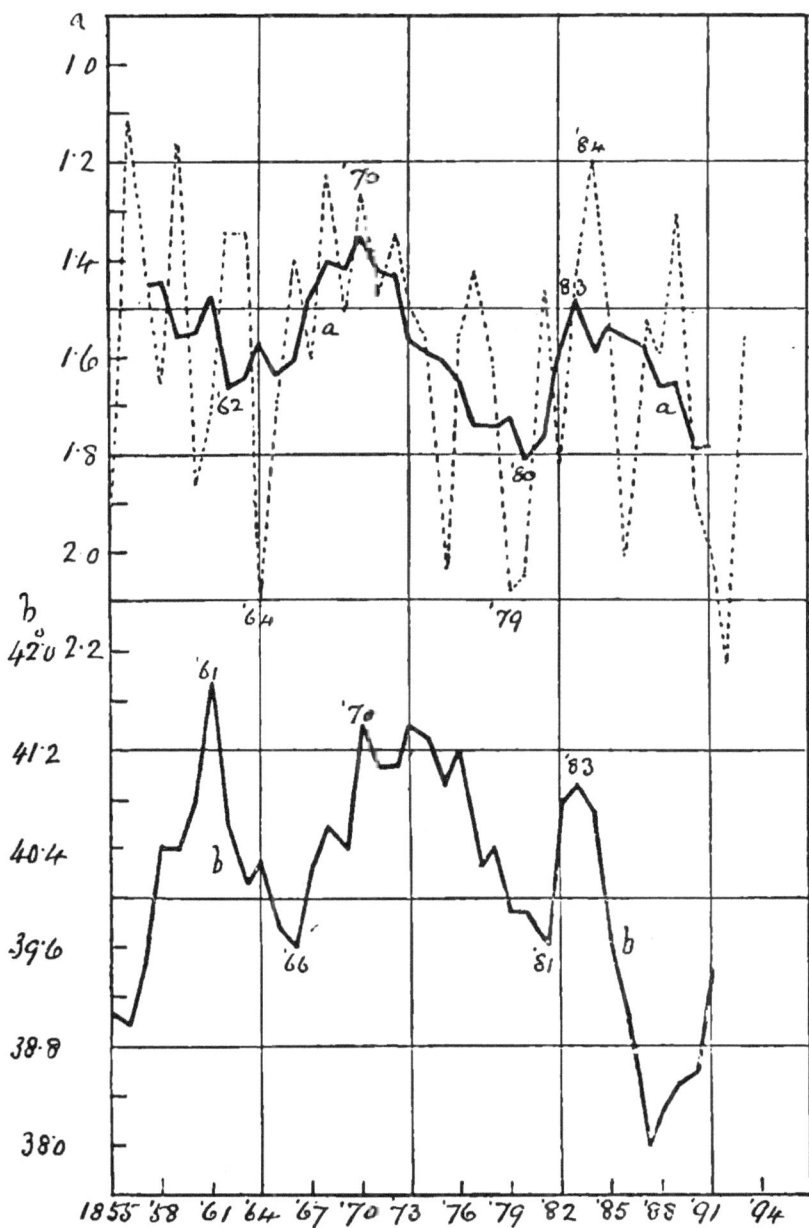

16.—ZYMOTIC DISEASE IN SEVERE AND MILD WINTERS.

DIAGRAM: Six smoothed curves (3 av.) of weekly deaths from zymotic disease in London, during first quarter of six years named. Continuous curves, severe winters ; dotted curves, mild winters.

The relation of zymotic diseases to cold is a somewhat complex subject. It may be well to ask, What kind of mortality has there been from such disease in recent severe winters ? It will be remembered that the winters 1878-79, 1890-91, and 1894-95 were all notably severe. Allowing for development of disease, we note the weekly deaths from zymotic diseases (as specified under 13), not in the 3 winter months, but in the first quarter, and smoothing the figures with averages of 3, we obtain the three continuous curves. Then we do the same with the three mild winters 1882-83, 1883-84, 1884-85 (dotted curves).

It may be noted that the three mild winter curves all show more deaths from zymotic disease than the curves of the very severe winters 1890-91 and 1894-95, notwithstanding growth of population and growth of at least diphtheria. The 1879 curve shows a higher mortality than those two, coming second to that of 1885; (there was a great deal of whooping cough). On the other hand, we must remember the progress that has taken place in overcoming zymotic disease generally, and of this the relative position of the curves 1879, 1891 and 1895 is no doubt partly an expression. Still, it appears from various evidence that intense winter cold is on the whole unfavourable to the mortality from zymotic disease, and mild winter weather favourable ; and in this we may find some justification of the saying, " A green Yule makes a fat churchyard," which, if we take the large mortality from respiratory diseases into account, and its rise in severe cold, is rather falsified. (*Weekly Return.*)

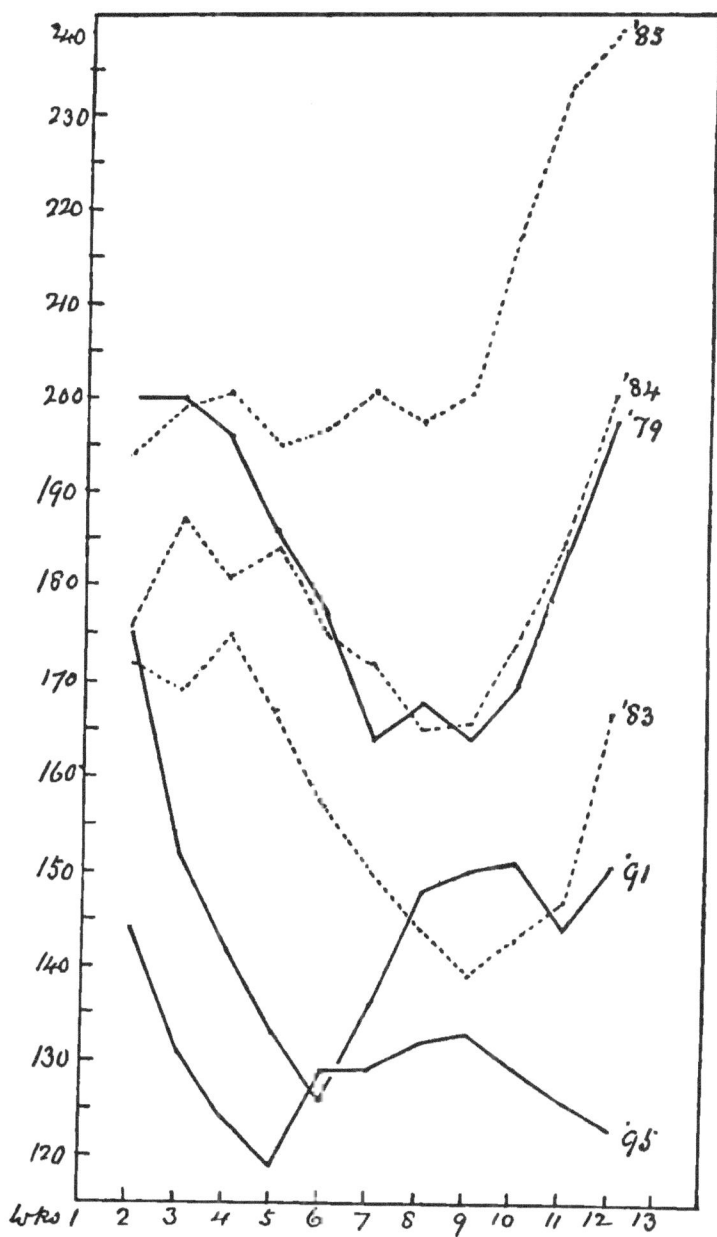

PRINTED BY

The Graphotone Co.,
BUSH HILL PARK,
ENFIELD.

www.ingramcontent.com/pod-product-compliance
Lightning Source LLC
Chambersburg PA
CBHW021426090426
42742CB00009B/1267